Urban Morphology and Sustainable Smart Cities

Gouri Sankar Bhunia · Pravat Kumar Shit

Urban Morphology and Sustainable Smart Cities

Application of Geospatial Technology

Springer

Gouri Sankar Bhunia
Independent Researcher
Kharagpur, West Bengal, India

Pravat Kumar Shit
PG Department of Geography
Raja Narendralal Khan Women's College
Midnapore, West Bengal, India

ISBN 978-3-031-97472-4 ISBN 978-3-031-97473-1 (eBook)
https://doi.org/10.1007/978-3-031-97473-1

This Springer imprint is published by the registered company Springer Nature Switzerland AG
The registered company address is: Gewerbestrasse 11, 6330 Cham, Switzerland

Preface

In recent times, urban areas around the world have faced a significant challenge as countries transition towards market-based economies and uphold the principles of subsidiarity outlined in their respective constitutions. Municipal authorities are now creating a third tier in the national policy framework. In the light of this ongoing and future development, many countries globally have introduced various programmes aimed at strengthening and improving municipal services. Unfortunately, due to inadequate planning and governance in peri-urban areas by local governments, numerous difficulties have emerged. The concept of a "Smart City" offers a technologically advanced approach to address issues stemming from unplanned urbanization, with a focus on delivering better citizen services and promoting sustainable practices. A "Smart City" is well-suited to managing and addressing technical, social, and economic development challenges. It necessitates specific foundational features such as cultural and physical infrastructure designed to foster competitiveness, collaboration, technological innovations, and the groundwork for new enterprise formation. Geospatial technology plays a pivotal role at every stage of planning, modelling, management, and expansion of smart cities, encompassing a wide range of functions. This brief underscores the importance of national governments and policymakers worldwide formulating comprehensive policies to ensure the planned and sustainable growth of urban and town areas within their respective countries.

This book provides an in-depth exploration of the challenges arising from the rapid worldwide urbanization. Unplanned urbanization has resulted in altered urban land use patterns, the rise of slums, subpar housing, hygienic problems, extensive pollution, and the propagation of worldwide diseases. It places a particular emphasis on stratagems and amenities aimed at addressing the problems and concerns stemming from urbanization, largely driven by rural residents seeking employment and improved living conditions in urban areas. Furthermore, the book offers comprehensive insights into how contemporary urban challenges are being tackled by professionals in fields such as design, architecture, urban planning, and landscape architecture. These difficulties include problems with relocation, climate change, adaptability, the political process, and deterioration of the environment.

The contributors to the book transcend technical terminologies to show the political, social, and physical implications of digitizing the world in the framework of smart cities. They do this by drawing on backgrounds spanning environmental design, geography, organizational development, scientific and technological uses, and engineering and architectural design. Along with offering concise examples of urban innovation, environmentally friendly remedies, and the long-term sustainable expansion of smart cities globally, the book also examines how to improve the state of the art and practice in the establishment of smart cities globally. In addition to covering real-time management, the Internet of Things, and the structure of systems for critical services, the book focuses on the use of geospatial technology in smart cities. In order to properly manage smart cities and promote a healthy urban environment, it explores land use planning, economic growth, and ecosystem-based understanding. The book examines the divergent facets of research on smart cities and provides insightful information on how policies are developed and put into practice to improve the procedures by which decisions are made in smart cities. Additionally, it provides readers with a deeper understanding of both the theoretical and practical aspects of smart city development. This book will prove invaluable to urban land managers, urban designers, those involved in planning and development, environmental management, economics, and geography courses, as well as property practitioners, catering to both undergraduate and postgraduate students.

Kharagpur, India Gouri Sankar Bhunia
Midnapore, India Pravat Kumar Shit

Contents

About the Authors

Gouri Sankar Bhunia has received a Ph.D. from the University of Calcutta in West Bengal India in 2015. He is working as Remote Sensing and GIS Expert since last 12 years in various government and private organizations. His Ph.D. dissertation work focused on disease transmission modelling using geospatial technology. Dr. Bhunia currently involved as GIS Expert in various smart city implementation programmes in India. His research interests include urban planning, infrastructure planning and mapping, environmental modelling, risk assessment, natural resources mapping and modelling, data mining, and information retrieval using geospatial technology. Dr. Bhunia awarded Senior Research Fellowship Grant from Indian Council of Medical Research. Dr. Bhunia is also involved as Visiting Faculty of various private universities in India. Dr. Bhunia is associated with the editorial boards of four international journal in health GIS and geosciences. Dr. Bhunia has published more than 80 articles and ten (10) books in an international journal in Scopus indexed and Publishers.

Pravat Kumar Shit (Ph.D.) is Assistant Professor at the PG Department of Geography, Raja N. L. Khan Women's College (Autonomous), West Bengal, India. He received his M.Sc. and Ph.D. degrees in geography from Vidyasagar University and PG diploma in remote sensing and GIS from Sambalpur University. His research interests include applied geomorphology, soil erosion, groundwater, forest resources, wetland ecosystem, environmental contaminants and pollution, and natural resources mapping and modelling. He has published twenty-three books (Springer-18, Elsevier-04, CRC Press-01) and more than 95 papers in peer-reviewed journals and 85 book chapters. He is also Guest Editor of *Environmental Science and Pollution Research* and *Applied Water Science.* He is currently Editor of the *GIScience and Geoenvironmental Modelling (GGM)* Book Series, Springer-Nature.

List of Figures

List of Tables

Chapter 1
Basic Introduction of Smart City

1.1 What is Smart City

Today, the nomenclature "smart city" is perhaps the most commonly used and embraced by people, the mass media, stockholders and businesses, and government officials. It is used by government agencies and the private sector because it is a buzzword that much of the target audience understands (Höjer and Wangel 2015). Information and communication technology, or ICT, is used in smart cities to improve citizen welfare, assistance from the government, and operational efficiency. Currently, 54% of the world's population lives in cities. Estimates indicate that about 70% of people on Earth will reside in cities by the year 2050. Due to population growth, urbanization will result in the addition of 2.5 billion people living in cities over the course of three decades from now. Although some megacities are still struggling to cope with the current influx of inhabitants, smart cities are needed to make these cities more liveable and fully sustainable (Pellicer et al. 2013). The main goal of a smart city is to maximize public services, promote economic growth, and improve the quality of life for its citizens by utilizing smart technologies and data gathering (Fig. 1.1). Technology is only as important as its applications, not as much as it is affordable.

The smartness of a city is governed by a range of factors, such as:

- An infrastructure focused on technology.
- Environmental protection initiatives.
- Effective and reliable public transport, and bold.
- Innovative town planning strategies.
- Those who will reside, labour, and benefit from the city's prosperity.
- According to Singapore Smart Nation, "smartness" is not a degree of how advanced or complex the technology being adopted is, but how well a society uses technology to solve its problems and address existential challenges."

G. S. Bhunia and P. K. Shit, *Urban Morphology and Sustainable Smart Cities*,
https://doi.org/10.1007/978-3-031-97473-1_1

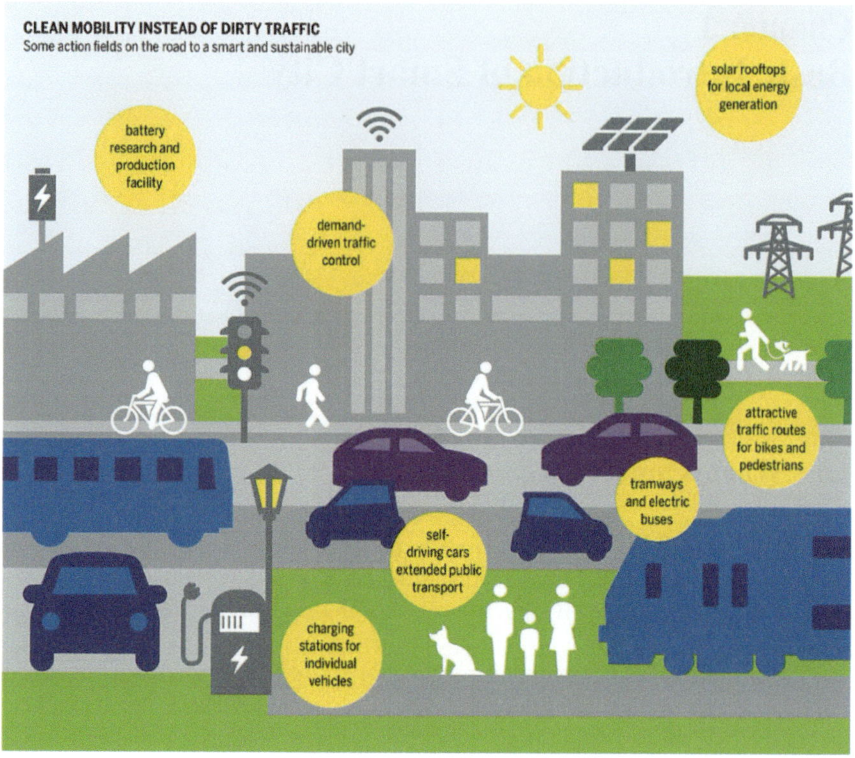

Fig. 1.1 Possible scenario of smart and sustainable mobility

- Business Dictionary—"A smart city is a developed urban area that creates sustainable economic development and high quality of life by excelling in multiple key areas; economy, mobility, environment, people, living, and government."
- Boyd Cohen—"A broad, integrated approach to improving the efficiency of city operations, the quality of life for its citizens, and growing the local economy."
- IoT World—"A smart city is a municipality that uses information and communication technologies to increase operational efficiency, share information with the public and improve both the quality of government services and citizen welfare."
- According to Deakin and Al Waer (2011), there are four factors that go into defining a smart city:

  i. The application of numerous technological and cutting-edge technologies for the good of towns and communities.
  ii. The application of ICT to change the region's living and working environments.
  iii. The incorporation of these information and communication technologies (ICTs) into administration structures.

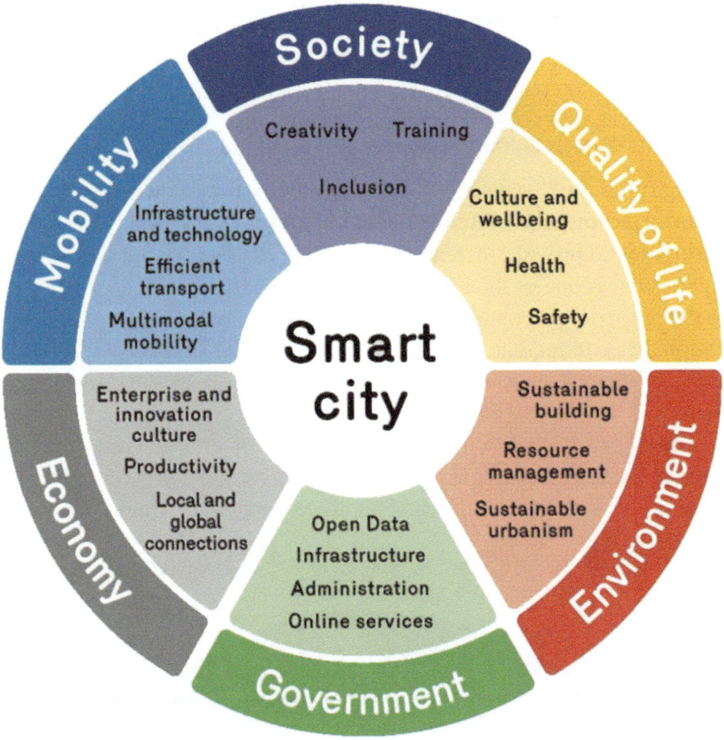

Fig. 1.2 Area of smart city model

iv. The territorialization of activities that put ICTs and people collected to improve the level of creativity and expertise available.

- "The term is not static; there is no absolute epitome of a smart city, no end point, but rather a phase, or sequence of steps, through which cities become more "liveable" and resilient, and therefore more able to adapt more quickly to new challenges," as stated by the UK Department for Business, Innovation and Skills in 2013.
- "Smart cities deliver prosperity in terms of economic practices and employment opportunities to a large section of their citizens, irrespective of the level of education, skills, or income levels," according to the Indian government in 2014 (Fig. 1.2).

1.2 Necessity of Smart City

Currently, 54% of the world's population lives in cities, and by 2050, that percentage is expected to climb to 66%. Over the next three decades, urbanization will bring another 2.5 billion inhabitants to cities due to population growth (Patrão et al. 2020).

Sustainability in terms of the environment, society, and economy is required for maintaining with the fast growth that is straining the resources of our cities. At the September 2015 United Nations meeting, 189 countries agreed on the sustainable development goals (SDGs) programme. Time is crucial because, as we all know, centralized choices and implementation might require an extended period of time.

- The primary goal of a smart city is to create an urban environment that raises government income while offering people a good level of living. Thus, the capacity of smart cities to provide increased quality services to people with fewer assets and investment is one of their most important benefits.
- Cities that experience population growth must optimize their resource and asset utilization to accommodate the expanding population. Smart city apps will help to make these changes, as well as enhance business processes and residents' living standards.
- Smart city apps will help cities make the most of their existing infrastructure by helping them find new possibilities. By enabling new revenue streams and operational savings, the improvements save money for both individuals and governments (Fig. 1.3).

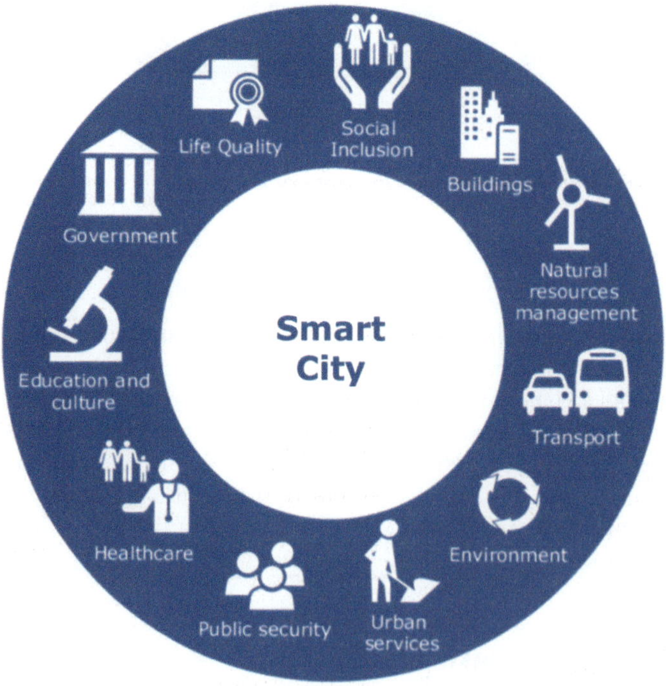

Fig. 1.3 Facilitate of urban areas into smart cities (modified after Lavalle et al. 2020)

There are numerous opportunities to implement initiatives that facilitate the transformation of urban areas into smart cities. However, this shift does not occur automatically; it requires deliberate and informed decision-making by city officials. In this context, data and AI-generated insights may hold little value unless they are effectively represented. To support urban policymakers in making well-informed choices for the sustainable development of smart cities, it is crucial to present data in a clear and comprehensive manner. Ideally, this should be achieved through impactful visualizations that drive actionable change. Nevertheless, the vast amount and complexity of data generated by smart city technologies pose significant challenges in achieving this goal.

1.3 Components of Smart City

A smart city roadmap has four/five main elements:

- Figure out exactly what a society is: it has to do with natural features, acquaintances between urban and rural areas, and people movements between them; it is also possible that in certain countries, the given concept of city/community does not correlate efficiently to what essentially happens in real life.
- Evaluate the community: This can be achieved by figuring out the benefits of that kind of intervention. Investigate the neighbourhood to find out about its residents' distinctive qualities, such as age, level of education, interests, and city facilities, as well as the community's members and expectations for customers.
- Opening the data vault: Governments, entrepreneurs, and individuals have traditionally kept their data close to their pockets, exposing it with as few people as possible. Even so, all actors in the dynamic ecosystem exchange knowledge and integrate it with circumstantial information gathered in real-time, which is a crucial enabler of supportable smart cities. This is how real-time, well-informed decisions are taken.
- Develop an effective city policy: Identify roles, duties, objectives, and purposes for establishing a policy that will direct the development projects. Establish strategies and strategies to meet the goals.
- Engaging people: This can be proficient by using e-government programmes, open data, sporting activities, and other methods to engage citizens.
- Digital government: The objectives of digital government are to make community facilities more accessible by distributing them exclusively to a citizen's laptop or smartphone screen. A digital government encourages people and states to communicate and collaborate with one another, while also lowering the cost of public facilities and distributing them in a more timely and safe manner (Fig. 1.4; Table 1.1).

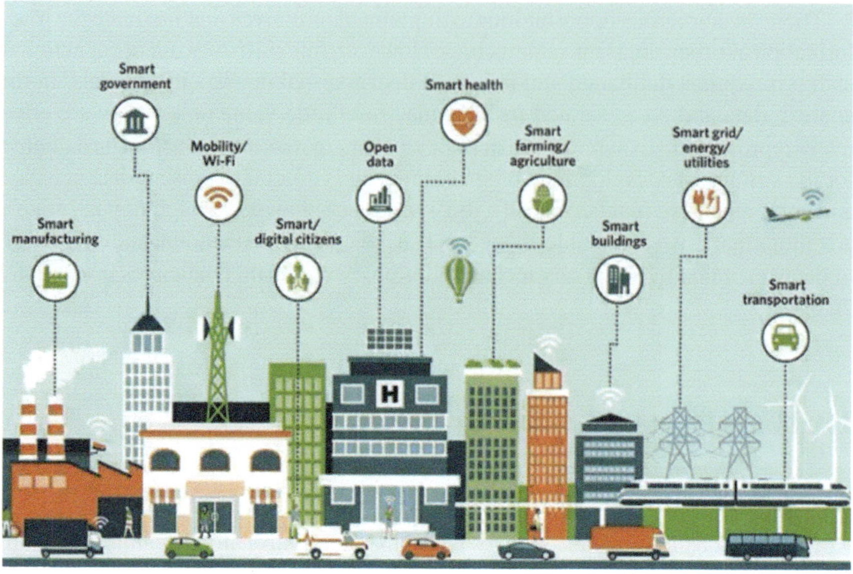

Fig. 1.4 Smart city elements

Table 1.1 Main areas of intelligent city activation

Innovation economy	Urban infrastructure	Governance
Modernization in industries, bunches, region of a city	Transport, buildings, manufacturing	Administration facilities to the inhabitant
Knowledge workforce: Education and employment	Energy/utilities, farming, health care	Participatory and direct consensus
Formation of knowledge-intensive businesses	Fortification of the atmosphere/security, solar energy	Facilities to the citizen: Superiority of life

1.4 History of Smart City

Considered the "Father" of the planning of cities and a pioneering Greek city planner, Hippodamus of Miletus was an ancient Greek architect, weather forecasters, mathematical analyst, and philosopher. The need to restore cities and nearby region was enormous after decades of war, ruin, and colonization of his homeland. His idea was centred on the separation of cities into sections for private, public, and religious use. With today's greater cities and transportation demands, developers prefer mixed-use neighbourhoods over sharp divides. Miletus is in charge of laying out the city planning grid system. His vision and arrangement applied in the reconstruction of Ancient Greek cities can be specifically traced to today's wide boulevards interconnecting at right angles and city centres.

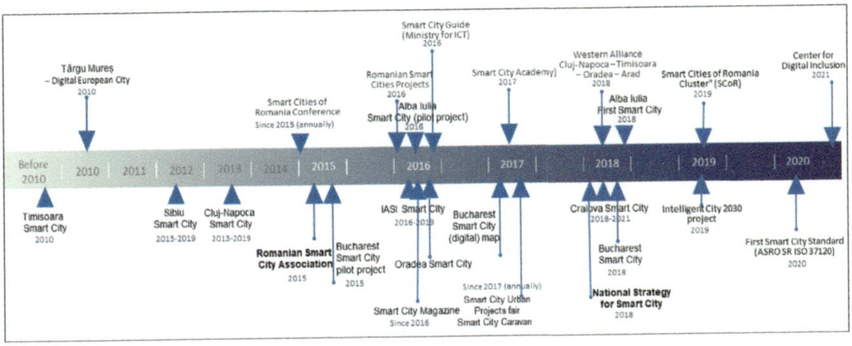

Fig. 1.5 Example of the evolution of Romanian smart cities (modified after Ibanescu et al. 2022)

The idea of smart cities dates back to the 1960s and 1970s, when the US Community Research Bureau started collecting data, directing money, and delivering intelligences to guide utilities, alleviate floods, and minimize suffering using datasets, aerial imagery, and cluster analysis. Subsequently, smart cities have evolved into three separate generations.

- The initial wave of smart cities was provided by technology suppliers so that people might better understand how technology affects daily living.
- This gave rise to the second wave of smart cities, which looked at how other technological advancements and smart devices could be combined to create comprehensive municipal services.
- Public participation was included in the third wave of smart city models, which delegated responsibility from local governments and technology providers to citizens and participation from the community.

Vienna followed this third-generation approach, forming an alliance with the local Wien Energy provider to encourage people to participate in local solar plants while also partnering with the public to address concerns for instance gender equity and reasonable housing. Adoption of this kind has continued throughout the world, notably in Vancouver, where 30,000 people collaborated to establish the Vancouver Greenest City 2020 Action Plan (Fig. 1.5).

1.5 Smart City—Global Scenario

Smart economy, smart sustainable policies, smart government, smart living, smart mobility, and smart citizens are the six major groups that make up the Smart Cities Wheel (Joss et al. 2019). North America, Europe, Asia/Pacific, and Latin America were divided into four regions (https://www.un.org/). The findings revealed that European cities are the world's smartest; they are denser, have more public transportation, a greater dedication to pedalling and rambling, and a greater prominence on recycling and low-carbon strategies. Europe's top city was Copenhagen, which placed above Amsterdam, Vienna, Barcelona, Paris, Stockholm, London, Hamburg, Berlin, and Helsinki.

- In terms of digital administration and open data, Seoul is the smartest capital. With more than 1200 publicly available data sets at its disposal, the city has paved the way in utilizing technology to foster citizen involvement. One such example is the OASIS Online Policy recommendation System, which invites residents to provide online suggestions for policymaking.
- Singapore is a tidy, orderly, technologically sophisticated community with outstanding public transportation and a firm dedication to long-term sustainability. The city has one of the smallest carbon emissions of any big city in the country, emitting only around 2.7 tonnes of CO_2 per capita.
- Tokyo Metropolitan Government was the primary town in the ecosphere to implement a city-based cap and trade scheme, which is a pollution reduction programme that limits gross emissions of such pollutants while allowing emitters to trade usable permits on an open market as part of enforcement action.
- Hong Kong is regarded as one of the world's most creative metropolises. The Hong Kong Innovation and Technology Commission, which was established in 2000, helped to fund the development of five research "clusters" focused on automobile, ICT, logistics, nanotechnologies, and textiles. The commission has helped to generate more than $3 billion in research and development investments around the five focus clusters since its inception.
- Auckland has long been observed as one of the most liveable capitals in the world, and it is home to two-thirds of the country's top 200 enterprises. The city pledged to becoming one of the first nine cities to collaborate with Microsoft in the introduction of their CityNext app, which seeks to improve local government and promote creativity through the use of ICT and big data in fields like transportation, energy use, and buildings.
- Sydney took advantage of the money and publicity that came with hosting the Olympic Games to green the area. The 90-hectare Sydney Olympic Village is a mixed-use construction effort that features renewable energy technologies like 12 solar panels per house, a constructional waste recycling programme that recycles up to 90% of waste brick, pavement and lumber, and, of course, improved public transport connections.
- Melbourne has also taken strides in becoming a more environmentally friendly and intelligent community. The city set a lofty target of halving carbon dioxide emissions by 100% from 2006 levels. The city launched one of the largest urban solar programmes on the top of Queen Victoria Market in 2003, and in 2010, it launched the 1200 structures programme to make it possible to adapt the city's massive structures.
- Since 2011, in addition to improvements in living standards and transportation, Osaka has been researching with smart house technology. The city's smart home initiative used a mix of renewable energy technologies and a home energy management system (HEMS) in partnership with Osaka Gas, resulting in an 88% reduction in electricity usage compared to equivalent homes in Osaka.
- Establishing inventory in Kobe has been made more environmentally friendly through the efforts of the Comprehensive Assessment System for Built Environment Efficiency (CASBEE) Programme. Via the CASBEE initiative, Kobe

currently has 450 green-certified houses. The city has since attempted to change its urban water and waste management activities over the last decade, changing its waste-to-landfill policy to waste and wastewater collection and reuse.

- Perth Solar City, a three-year initiative, started in 2009 with the goal of encouraging industry, companies, governments, and the general communal to transform the technique they generate, consume, and conserve electricity. People who took part in Perth Solar City saved a total of $1 million in electricity in 2012.

1.6 Smart City—Indian Scenario

Countries all over the universe are endeavouring to brand their cities smarter by combining current and emerging technologies to accomplish a variety of goals more successfully. Some cities are being erected from the ground up to be smart, in order to provide people with cost-effective services while also being environmentally friendly, financially vibrant, and enticing. At 6.4% of the worldwide gross domestic product (GDP), India's economy ranks third globally in terms of purchasing power parity (PPP). With over 1.2 billion inhabitants, the country is also the second-largest in terms of population, with almost one-third of the population living in cities. Table 1.2 shows the trend of urbanization in India. The country's urban population grew from 17.3% in 1951 to 31.2% in 2011. Indian cities have experienced rapid urbanization over the last decade, with Delhi leading the way with a 4.1% growth rate. The growth of smart cities is hampered by our large population, unauthorized residents, illegal tenements, and extra-legal clients. Urbanization has brought together people from diverse social, cultural, economic, and philosophical backgrounds, which has led to a rise in anxiety, disputes, and corruption. Assault on women has become a big issue. As a result, Indian cities do not serve as models for planned, inclusive, stable, and long-term growth. The current state of affairs may be brought about by the fact that many urban areas are still primarily uncontrolled and unapproved, which leads to a high volume of people using public spaces and driving in an unorganized way that is representative of the unregulated economy, commercial enterprises, and consumers. A closer examination reveals significant disparities in physical growth and basic infrastructure and amenities within and within cities. The wealthiest individuals live in exclusive neighbourhoods with excellent planning and services, whereas low-income families reside in slums and informal settlements with limited or no access to basic amenities. When it rains, water logging occurs in many areas, limiting mobility even further.

Around 843 million people are anticipated to animate in Indian cities by 2050. India must find smarter ways to enhance the quality of life as a result of its rapid urbanization (Ritchie and Roser 2018). According to experts, about 25–30 individuals will commute from rural sections to major Indian cities every minute in seeking a healthier life. As a result, India's cities must have twin objectives. To begin with, model new cities should strive to be economic powerhouses with high living standards and long-term environmental sustainability. Second, older cities can gradually become wealthier, more effective, and more sustainable. Both of these objectives can be difficult to achieve due to financial and managerial limitations. Creating a couple

Table 1.2 Urbanization in India

Year	% of urban population	No. of towns	Total population (millions)	Urban population (millions)
1901	10.8	1827	238.39	25.85
1911	10.3	1825	259.09	25.95
1921	11.2	1949	251.32	28.09
1931	12.0	2072	278.98	33.46
1941	13.9	2250	318.66	44.16
1951	17.3	2843	361.23	62.44
1961	18	2365	439.23	78.13
1971	19.9	2590	548.15	109.11
1981	23.3	3378	159.46	159.56
1991	25.7	3762	846.30	217.61
2001	27.8	4368	1048.15	296.97
2011	31.16	7935	1210.20	377
2021 (Estimated)	35.39	–	4981.79	475

Source Census of India

of smart city models that fulfil all of these targets and following up will be a viable option.

2015 witnessed the inauguration of the "100 Smart Cities Mission" by the Indian government (Government of India, 2014). Among other things, these cities were meant to feature more open areas, more housing for everyone, and improved infrastructure. The mission essentially calls for formulating and applying smart solutions to overcome the challenges confronting various urban sectors, such as water supply, sanitation electricity, mobility, housing energy, and environment (MoUD 2015). Figure 1.6 illustrates the smart city mission process in India.

1.7 Future Smart City

Future smart cities would emerge as a consequence of the fusion of numerous new generation innovations and remedies across all spheres of society, growing, changing over time, and producing new objectives and obstacles. As long as new difficulties are overcome and goals are realized through new technology, cities and civilizations will continue to be intelligent (Table 1.3). It is possible to combine modern technology with the city's current infrastructure in a variety of ways, creating matrices that help to better understand their interdependencies. Once created, such matrices would undoubtedly aid in understanding the complex interplay of numerous socio-technical aspects and associated difficulties for actually executing plans.

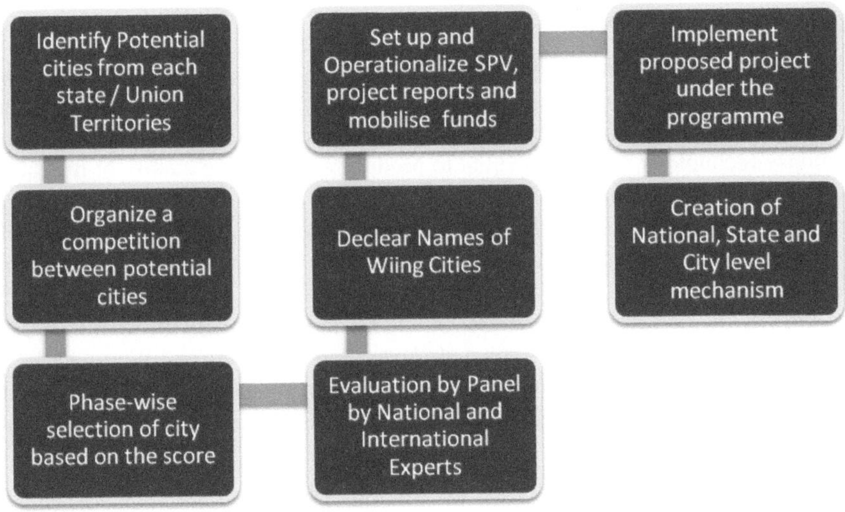

Fig. 1.6 Smart city mission process in India

Table 1.3 Various ideas of future smart city

Physical	Social	Economic	Governing
Digital	Inclusive and Integrated	Business-friendly	Efficient
Eco-friendly	Safe and secure	Competitive	Law abiding
Green	Educated and healthy	Entrepreneurial	Managed
Resilient	Intelligent and participative	Innovative and productive	Productive
Sustainable	Good amenities (air, water, food, sanitation, medical)	Productive and resilient	Well-run, Well-led

1.8 Smart City Challenges and Concerns

Urban digital transformation is beset with challenges, even with the seeming benefits of smart city development. Here is a rundown of some of the most important issues impeding the growth of smart cities:

- **Economic challenges**

Technology and ICT foundations are necessary for the development of smart cities (Hollands 2008); they are practical (IBM 2012) and advantageous in the efficient use of resources, the reduction of pollution, the cost of travel, the competitiveness of the cities, the growth of GDP and tourism, and the advancement of the economy. More than 50% of manageable trash and 70% of energy savings have been documented for climate street (Deakin 2013). But building smart cities need funding and help from both domestic and foreign governments.

- **Technology challenge**

Smart cities demand smart technology in the areas of solar cells, routers, transparency and urban information dissemination, online banking and welfare, etc. (Correia and Wünstel 2011). These amenities are either not there or of inadequate quality in several nations. Technology has the potential to produce and improve quality of life, but it also raises the possibility of economic inequality (Washburn et al. 2010). In several cities, the advantages of IT are still not evident (Chourabi et al. 2012). As a result, city administrators need take into account a number of aspects while implementing ICT because of the institutional capability, the resources available, as well as the disparities, digital divide, and cultural shifts (Washburn et al. 2010). Additional ICT issues are technical and security-related. According to Correia and Wünstel (2011), a smart city requires data security (encryption, authentication, and authorization), device connection and engagement, power for sensors, end-user interaction, and service discovery.

- **Infrastructure**: The difficulties of building high-speed Internet connections and renovating decades-old infrastructure, such as steam plumbing, underground cables, and highway tunnels and passageways continue to plague major metropolitan areas. Even though broadband wireless service is becoming more widely available, several areas of large cities still have inadequate connectivity. Funding for new developments in infrastructure is limited, and the approvals procedures can drag on for years. Because of the setting up of new sensors and other improvements, residents of these cities confront brief but stressful obstacles.
- **Security and privacy**: After learning about the threats posed by cyber terrorism to vulnerable and outdated power grids, everyone seemed to be more conscious of and sceptical of the relationship between security and technology. More money and resources are being allocated by smart cities to defence, and IT companies are creating novel technologies with integrated security measures against cybercrime and cyberattacks. The increased interconnectivity of mutually interdependent silos of city infrastructure, as well as the increasing number of IoT sensors, poses legitimate concerns. If security protocols do not improve, cybercriminals will one day be able to bring down an entire city. Blockchain technology is currently the hottest thing in technology; thus, many developers are trying to figure out how to incorporate these safe protocols into new apps to increase security.

Surveillance cameras mounted on every city corner can deter speeding and other offences, but understanding that "big man" is watching may make people feel a variety of emotions, from uncomfortable to paranoid.

- **Educating and engaging the community**: In order for a smart city to thrive, it requires "smart" citizens that are dedicated to utilizing modern technologies. Community learning is an essential component of the enactment process for every innovative citywide technology effort. People who feel that they have a voice in important decisions that affect their daily lives and who receive their information in an open and thoughtful manner are more inclined to use technology and encourage others to do the same. For the creation of a smart city, this is essential.

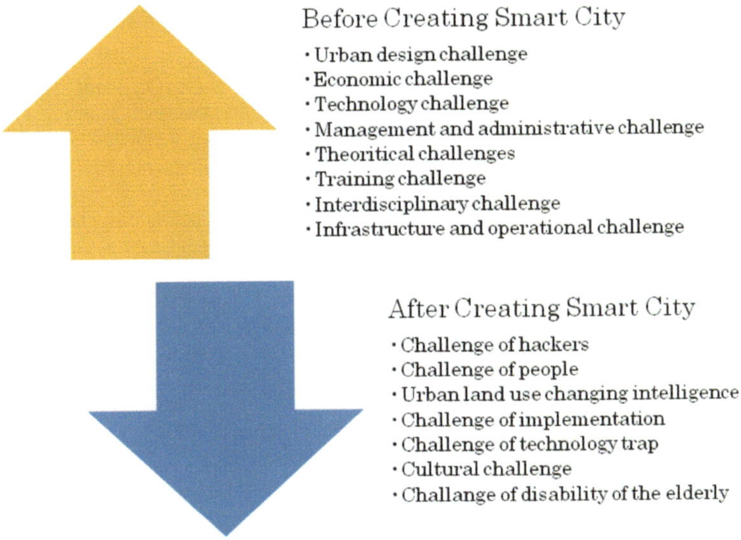

Fig. 1.7 Mapping of smart city challenge

- **Social inclusion**: It is critical that planning for smart cities considers the needs of all citizens, not just the affluent and highly tech-savvy. However, technology ought to constantly be promoting human growth rather than creating new divisions among people according to factors like wealth or level of education.

For instance, a city may be unable to launch a healthcare programme for the elderly because the majority of them are unable to use technology. As a result, smart city policies should be introduced in a way that promotes social equality and speaks to all people, not just the wealthy and technologically savvy (Fig. 1.7).

- **Management and administrative challenge**

Some specialists bring up the topic of administering a smart city and its numerous facets (Madakam 2014). Public engagement and political decentralization are either non-existent or at rudimentary levels in various countries. The central government should prioritize public involvement and citizen empowerment in order to begin the process of creating a smart city (Dadkhah and Shahbazi 2015). In reality, the benefits of smart cities include access to information, open data, transparency, and informing the public. These benefits are necessary for improved collaboration and consolidation between the federal, local, and citizenry (Rodríguez Bolívar 2015). Rebuilding government with innovative concepts to attain transparency, democracy, and accountability is made possible by smart technologies (Townsend 2013). Of fact, certain regimes may employ sophisticated technologies and instruments to further their totalitarian objectives and repress the populace (Greenfield 2013).

References

Chourabi H, Nam T, Walker S, Gil-Garcia JR, Mellouli S, Nahon K, Pardo TA, Scholl HJ (2012) Understanding smart cities: an integrative framework. Proceedings of the 45th Hawaii International Conference on System Sciences (HICSS), IEEE, pp 2289–2297

Correia LM, Wünstel K (2011) Smart cities applications and requirements. Net!Works European Technology Platform — Expert Working Group Report

Dadkhah K, Shahbazi, M (2015) Electronic smart cities: concepts, definitions, characteristics, and trends. In: Proceedings of the International Conference on eDemocracy & Open Government (CeDEM), pp 13–24

Deakin M, Al Waer H (2011) From intelligent to smart cities. J Intell Build Int 3(3):140–152. https://doi.org/10.1080/17508975.2011.586671. S2CID110580067

Deakin M (2013) Smart cities: governing, modelling and analyzing the transition. London: Routledge

Government of India (2014) Ministry of Urban Development. Archived from the original (PDF) on 3 February 2015. "Draft Concept Note on Smart City Scheme" (PDF)

Greenfield A (2013) Against the smart city. New York: Do Projects

Hollands RG (2008) Will the real smart city please stand up? Intelligent, progressive or entrepreneurial? City 12(3):303–320

Höjer M, Wangel J (2015) Smart sustainable cities: definition and challenges. In: Hilty LM, Aebischer B (eds) ICT innovations for sustainability. Advances in intelligent systems and computing, vol 310. Springer, Cham, Switzerland, pp 333–349

IBM (2012) Smarter cities for smarter growth (Report). Armonk, NY: IBM Global Business Services

Ibanescu D, Ploae V, Floarea A (2022) Smart city initiatives and urban resilience: a systematic literature review. Sustain 14(9):5327

Joss S, Sengers F, Schraven D, Caprotti F, Dayot Y (2019) The smart city as global discourse: storylines and critical junctures across 27 cities. J Urban Technol 26:1, 3–34. https://doi.org/10.1080/10630732.2018.1558387

Lavalle C, Barredo JI, Camia A (2020) Smart cities and sustainability: a European perspective. J Urban Plan Dev 146(4):04020045

Madakam S (2014) Internet of things: smart things. Int J Futur Comput Commun 3(3):169–173

Ministry of Urban Development (MoUD), Government of India (2015) Smart cities mission statement & guidelines. New Delhi: Government of India

Patrão C, Moura P, de Almeida AT (2020) Review of smart city assessment tools. Smart Cities 3:1117–1132. https://doi.org/10.3390/smartcities3040055

Pellicer S, Santa G, Bleda AL, Maestre R, Jara AJ, Skarmeta AG (2013) A global perspective of smart cities: a survey. In: 2013 Seventh international conference on innovative mobile and internet services in ubiquitous computing, pp 439–444. https://doi.org/10.1109/IMIS.2013.79

Ritchie H, Roser M (2018) - "Urbanization". Published online at Our World In Data.org. Retrieved from: https://ourworldindata.org/urbanization [Online Resource]

Ritchie H, Roser M (2019) Urbanization. Our World in Data. Retrieved from: https://ourworldindata.org/urbanization

Rodríguez Bolívar MP (2015) Transforming City Governments for Successful Smart Cities. Public Administration and Information Technology book series, Vol 8, Springer

Townsend AM (2013) Smart cities: big data, civic hackers, and the quest for a New Utopia. New York: W.W. Norton & Company

Washburn D, Sindhu U, Balaouras S, Dines RA, Hayes NM, Nelson LE (2010) Helping CIOs understand "smart city" initiatives: defining the smart city, its drivers, and the role of the CIO. Forrester Research, Inc.

Chapter 2
Urban Space, Society, and Urban Innovation

2.1 Introduction

A public space is a place where people can congregate, connect, and express their culture and lifestyle (Kurniawati and Prihantini 2018). Democratization and collective needs in events are at the heart of European public space (Mumford 1966). Although the Indonesian public space is a product of the government's power and generosity towards its people (Kurniawati 2018), public spaces are undergoing significant alterations of a variety of policies, forces, and urban frameworks. The design of open space is shaped by economic, political, and technical factors. When the government wants to make neighbourhoods kid-friendly, they build playgrounds in all public spaces. New construction is proposed to substitute the services of properties that are considered transferable when the governing body backs a green city. Hotspots and Wi-Fi appear to be ubiquitous in smart city design. The idea behind conflicting policies is, of course, that they will support the proper use of public space.

2.2 Urban Space and Smart City

In the last two decades, public spaces have regained prominence in urban planning discourse as a necessary component of urban sustainability. Public spaces are crucial in the development of inclusive cultures, as well as in the enrichment of cultural diversity and the development of a cultural life (Low et al. 2005). Most of the discussion around smart cities revolves around how to digitize infrastructure using a variety of networking and Internet of Things (IoT) technologies. Moreover, the important concern of reinventing public space to contend with emerging tech-driven paradigms—such as the shared economy, innovation, and, more broadly, the new smart urban economy—has received relatively little attention (Radwan and Morsy 2018). From New Towns to eco-cities, new models of urbanism were physically

G. S. Bhunia and P. K. Shit, *Urban Morphology and Sustainable Smart Cities*,
https://doi.org/10.1007/978-3-031-97473-1_2

conceived as spaces and places for various types of urban activity (Caprotti 2018). The smart city is the latest evolution of a technologically advanced push to envision and form the city's and urban society's futures.

2.2.1 Technology-Driven Paradigms and Their Impact on Public Space

Urban city dwellers' lives are surrounded by social media and new technology. When technology advances at a breakneck pace, so do the apps and services that people use on a daily basis. Many of the new generation adolescents have already adopted social networking as a social standard in this technological society. Facebook status updates, WhatsApp, tweets, and Instagram photos have all become commonplace. Furthermore, economic activities have encouraged the utilization of public space while shopping was done at markets and retail establishments. Commercial activities have also encouraged the productive utilization of public space because marketplaces and retail establishments were the places to shop. The Internet has described "information agora" as a unique type of public space (Drohsel et al. 2010).While the use of digital networks becomes more commonplace, a new digital layer is applied to the current urban landscape, resulting in a multilayered view of the urban world through direct interaction and the implementation of digital technology.

Urban planners now have to deal with the role that these innovations play in how people utilize and communicate in public spaces as a result of the constant improvements in the realm of information, including the development of mobile phones, Internet access via wireless networks, Bluetooth, GPS, and all associated applications. These developments have changed how people interact with each other and their immediate environment.

2.2.2 Mobility

Two challenging concerns in many contemporary cities are public space and connectivity. These problems bring up important considerations, such what effect does public space have on the sustainability of urban mobility in today's cities? And what connection exists between improving mobility and living standards in public spaces?

A number of intercontinental establishments (OECD, EEA, UNCHS, CSD, World Bank, UN Habitat, and IISD) have shown a growing interest in developing benchmarks for assessing urban sustainability over the last two decades. Some of these groups have also established global monitoring systems for sustainable development. The vast majority of current and well-established indices of urban sustainability are composites of environmental, economic, and social factors. Several organizations and organizations have chosen various metrics based on their policy orientation and regional scope. As a result, the categories of "public spaces" and "mobility" are only considered in a restricted way within the current list of urban sustainability metrics and are only included in large categories like "open spaces," "urban quality," and "transportation." The most important current metrics related to mobility and public spaces are shown in Table 2.1. When looking at the Table 2.1, three things stand out. (a) Urban mobility measures are more prevalent than public space indicators. (b) A number of metrics prioritize the availability of parks and other green areas above streets, squares, and different kinds of public space that are more densely populated. It is well established that open and green spaces improve residents' well-being by addressing the values of comfort, accessibility, diversity of usage, and sociability. (c) Excepting those factors used by the World Bank, selected metrics seldom take into account such as environmental physical quality and the reasons why people want to use public spaces, walk, or ride. Despite the fact that these aspects are problematic to understand because of their idiosyncratic existence, we cannot overlook the fact that many components in the area, not only pedestrian roads or parks, appeal to a comprehensive evaluation of urban sustainability.

2.3 Case Study: GIS-Based Service Area Analysis for Educational Centre

Introduction

Every instance of human migration has temporal and spatial characteristics. For instance, depending on the goal of a travel, a person's residency may influence the choice of a destination, including school trips. Finding connections between the amount of time of education and labour market mobility has been the main goal of research on education and mobility, including studies on the relationship between schooling and earnings for men and women, the relationship between poverty and where people live, the globalization of education, and the consequence of school locations on the limited access that some social groups have to them.

Table 2.1 Public spaces and mobility indicators according to various sources (Ravazzoli and Torricelli 2017)

Organization	Mobility indicators	Public spaces
UNCHS	• Modal broke: the percentage of travel time spent at work by vehicle, rail, bus, bike, walk, or other • The average travel time to work, measured in minutes • Per capita spending on transport infrastructure • The quantity of cars per 1000 people	–
World Bank thirty-seven-city study (Newman and Kenworthy 1999)	• Journey to work (JtW) %: Private/Public/NMM (*4) • MJ/Pass Km: bus, rail, ferry • JtW Km • JtW mins • 100,000 people lose their lives in transit • Percentage of transported deaths overall • Total CO_2 per capita (kg) • NO_x per capita (kg) • SO per capita (kg) • CO per capita (kg) • VHC per capita (kg) • VP per capita (particles, KG) • Per capita road expenditure (USD) • %GRP allotted to transportation • Recuperated transit costs	–

(continued)

Table 2.1 (continued)

Organization	Mobility indicators	Public spaces
EEA	• Kilometres of carriageway and/or railway length • Number of average duration trips made per day per mode of conveyance per kilometre of population • Count of commuters entering and exiting the conurbation • Vehicle counts on major routes and the aggregate number of vehicles coming in and going out in km • Number of persons hospitalized in road accidents per 10,000 people	Accessibility of green space: percentage of people within 15 min walking distance of urban green areas
WHO healthy cities indicators (1994)	• Pedestrian streets • Cycling in the city • Living space • Public transport • Public transport network cover	• Public access to green spaces • Relative surface area of green space in the city
Urban audit Eurostat (2014)	• Automobile ownership: number of automobiles per 1000 population • Number of fatalities from traffic crashes per 10,000 residents • Percentage of users of public transient • Percentage of time spent riding and walking • The proportion of mass transient users	
Green City Index (EIU)	Modal split	The quantity of greenbelt areas, parks, and other open spaces
Alberti (1996)	Mobility and modal split	Sufficient amounts of public land and green places
UN-HABITAT (2016)		Both the quantity as well as the quality of public land

This study aims to assess the effectiveness of the GIS on the distribution of educational facilities in Kolkata city, in order to take advantage GIS technology to delineate the service area of the existing schools. The road network data and educational centre data of Kolkata city were collected from Open Street Map (OSM); after that, the geographical database was built. We analyse the service area of educational centre in terms of proximity to each other. Domicile enables the examination of accessibility inside a network based on service area analysis techniques, for example, what are the locations domicile can navigate to on a road network for less money (Fig. 2.1).

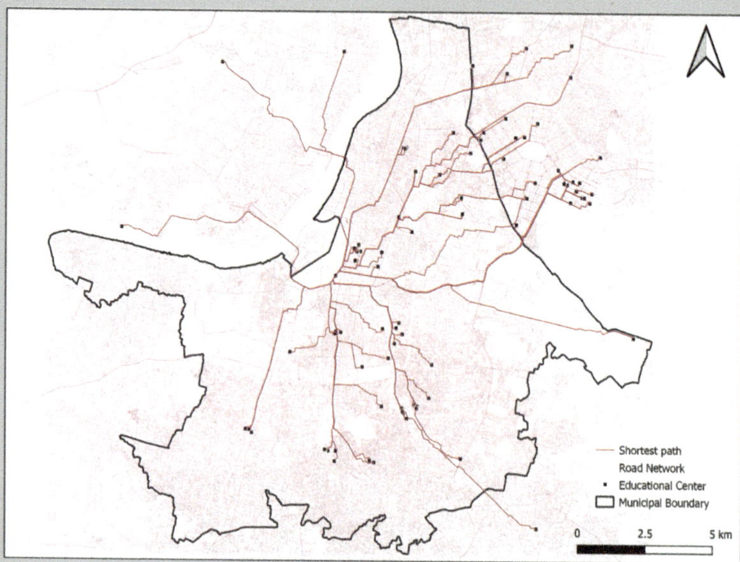

Fig. 2.1 Service area analysis of educational centre in Kolkata city

2.4 Urban Society and Smart City

Since the mid-twentieth century, our societies have been profoundly impacted by a number of worldwide ecological, social, and economic challenges (Yigitcanlar and Lee 2014).Urban centres have undertaken several initiatives to upgrade urban essential services, seeking at better environmental, social, and economic circumstances, enhance the attractiveness and competitiveness of cities, particularly during the last couple of decades (Jong et al. 2015).The intelligent use of digital data, for instance in the context of human health, mobility, energy use, education, transfer of knowledge

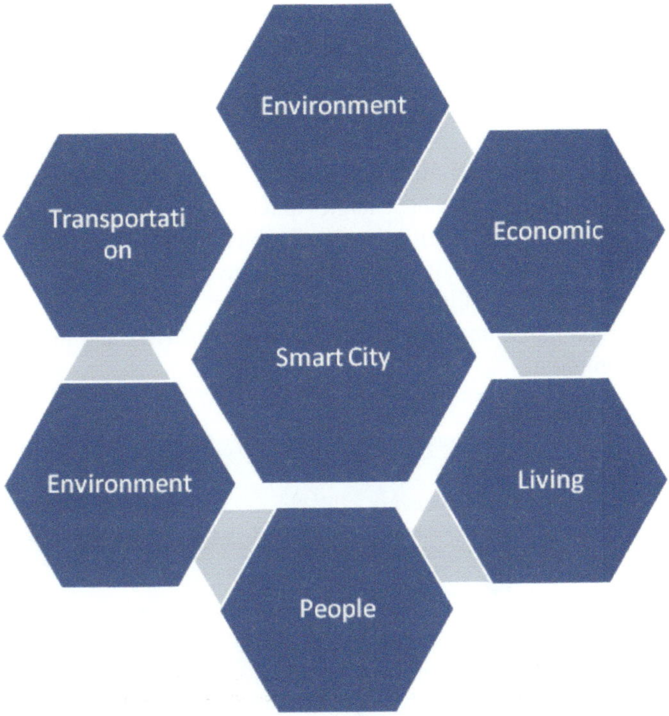

Fig. 2.2 General six dimensions model of smart city

and urban governance, has led to the emergence of smart cities, according to Deakin and Al Waer (2012) and Townsend (2013).

The term "smart city" refers to a city that is administered in a future-focused manner in the areas of economy, people, governance, mobility, and environment. It is also a city that is constructed autonomously, solely, and intentionally based on the intelligent amalgam of citizen operations. Six elements were afterwards used to define this idea: smart people, smart living, smart environment, smart transportation, smart economy, and smart governance (Giffinger 2007). By categorizing linked areas of application, including natural resources and energy, transportation and flexibility, buildings, life, governance, and the economy and people, an additional investigation looks at a comprehensive understanding of the notion of a smart city (Attaran et al. 2022) (Fig. 2.2).

2.4.1 *Smart Environment*

By fostering safe and healthy living circumstances, the smart environment aims to enhance citizens' quality of life (General et al. 2002). A smart energy distribution network, smart pollution monitoring and management, smart buildings, improving the integrity of natural resources, smart well-being, and smart promoting are a few of the significant components (Fig. 2.3).

Smart environments offer the potential for intelligent assessment and tracking, along with better control and surveillance of pollution in the city through technological innovations in order to optimize utilization of energy, the main goal of which is to encourage more use of renewable energies and less utilization of natural resources (Programme 2012). Smart surveillance of pollution mitigation may include tracking noise and climate pollution (Roy et al. 2018), waste management (Alqahtani et al. 2020; Idwan et al. 2020), building an economically viable green city with a view to lowering carbon emissions (Innovation and Carbon 2016), and managing urban water for consumption.

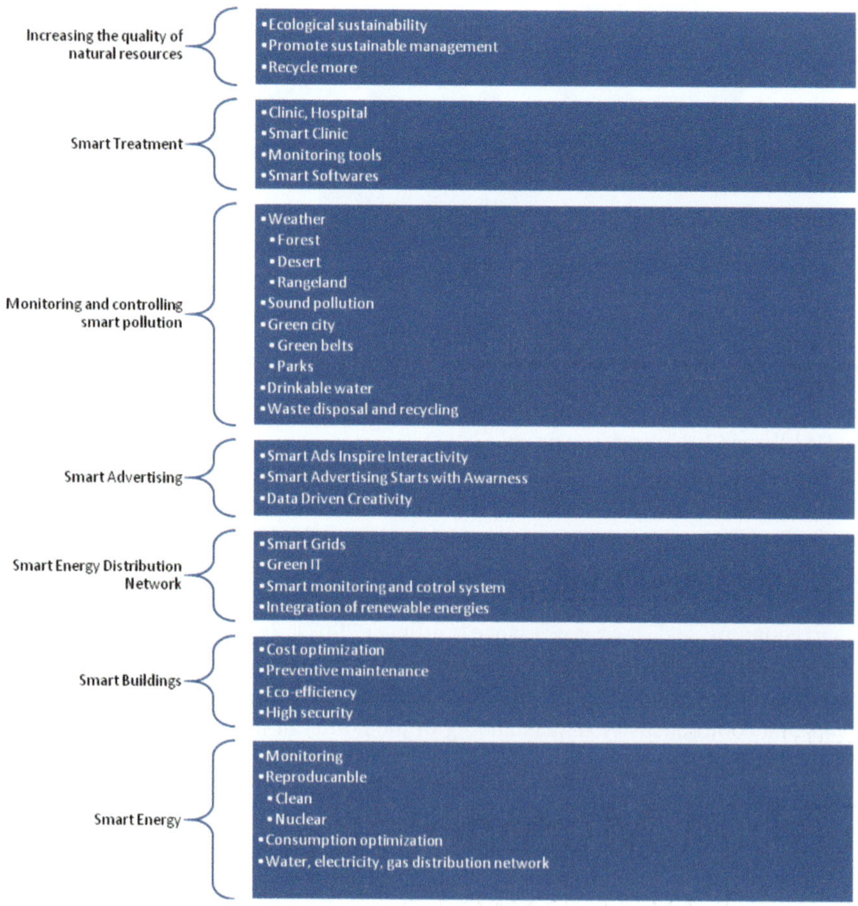

Fig. 2.3 Components of smart environment (modified after Attaran et al. 2022)

Smart energy also keeps an eye on how well-maintained smart water, electricity, and gas networks are, along with how to reduce energy consumption, in spite of controlling the usage of additional renewable resources like clean and nuclear energy (Villa-arrieta and Sumper 2019). It is vital to offer better services in the area of health due to the rise in urban population and the significant demand of individuals to discuss health. More effective communication between healthcare providers may save costs and increase access to patients' vital information. These are two benefits of using smarter methods to deliver medical services. By using the capabilities of the Internet of Things, an intelligent environment may make significant advancements in smart health. The Internet of Things constantly monitors vital signs of patients from a distance with the aid of tracking devices like attachment sensors, wearables, and biosensor sensors in the body. Using connected appliances, doctors can continuously track the well-being of patients so they can offer urgent care in a timely manner in case of emergencies (Moore and Van Pham 2019).

2.4.2 *Smart Economy*

The term "smart economy" describes a city's ability to compete on the basis of its creative use of new business models, e-commerce, inventive and creative product development, entrepreneurial activity, revolutionary ICT services, smart sales, smart industrial, smart agriculture, and smart tourism. By enhancing the business climate and increasing the city's allure for ability and investors looking to employ information and communication technologies (ICT) to grow the economy in an inventive way, the idea of a smart economy offers practical ways for bolstering the city's economy (Nin et al. 2019; Kumar and Dahiya 2017).

In order to establish a smart economy, which also involves a smart supply chain, smart advertising, marketing research, and market-making, smart sales, for instance, are one of the most crucial and essential aspects. Additionally, by using smart seeds, smart irrigation, and smart harvesting, smart farming has been able to use the Internet of Things to minimize energy consumption along with the costs connected to the use of manpower. This industry currently makes up a significant portion of global economic activity and is predicted to make up more than a third of all global trade in amenities (Al-omari and Alqaralleh 2020). Smart tourism generated by smart cities may boost the nation's economy. According to Nitti et al. (2017), "smart tourism" refers to the use of smartphone-based applications to deliver tourism amenities with the goal of giving visitors convenient access to tourist and religious sites as well as smart health centres. According to Zygiaris (2013), a smart economy is one where novel strategy for sustainable economic expansion, a consistent rise in gross national and domestic income, and an improvement in the trade balance are developed and put into practise using the resources that are accessible (Fig. 2.4).

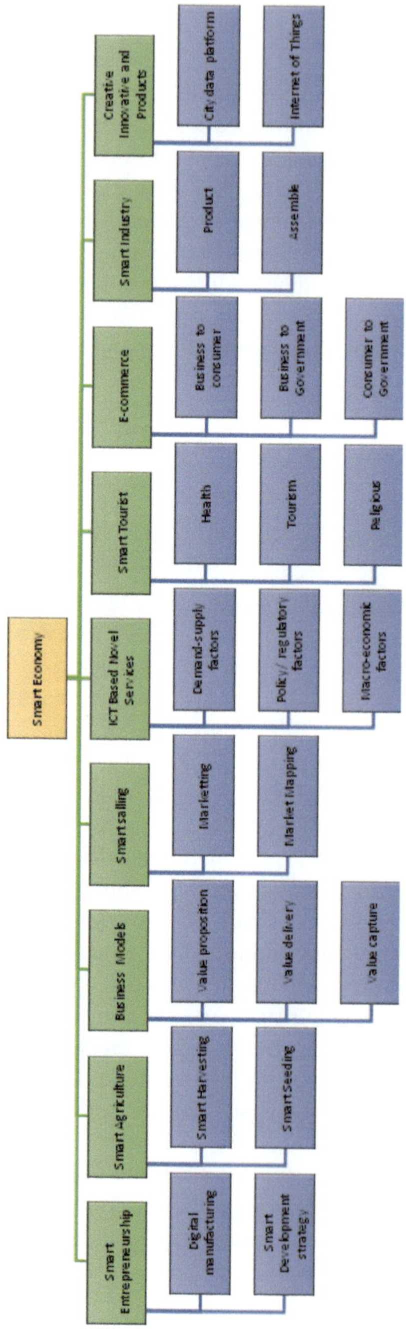

Fig. 2.4 Components of smart economy (modified after Attaran et al. 2022)

2.4.3 Smart Living

By offering secure and healthy living circumstances, the concept of smart living aims to improve residents' quality of life (General et al. 2002). The convenience of using numerous cultural and social amenities is one advantage of residing in smart cities. Living with a technologically based style, engaging in technologically based behaviours and uses, leading a healthy, safe, and revitalizing lifestyle, having access to a variety of cultural amenities in the city, and finding high-quality housing with intelligent solidarity are all examples of smart living (Fig. 2.5).

Although social cohesion is crucial for a city's overall growth, it is sometimes disregarded since many cities place a heavy emphasis on technological innovation. This element is determined by the degree of cohabitation in a city between various racial and socio-economic groups, as well as people of various ages, jobs, and cultures. The link between this subgroup and other subgroups is unavoidable since smart living necessitates a healthy environment, social and personal security, a sufficient income, and both individual and public health. Consequently, it is clear and demonstrated that this subgroup has a link with other subgroups in the form of a graph.

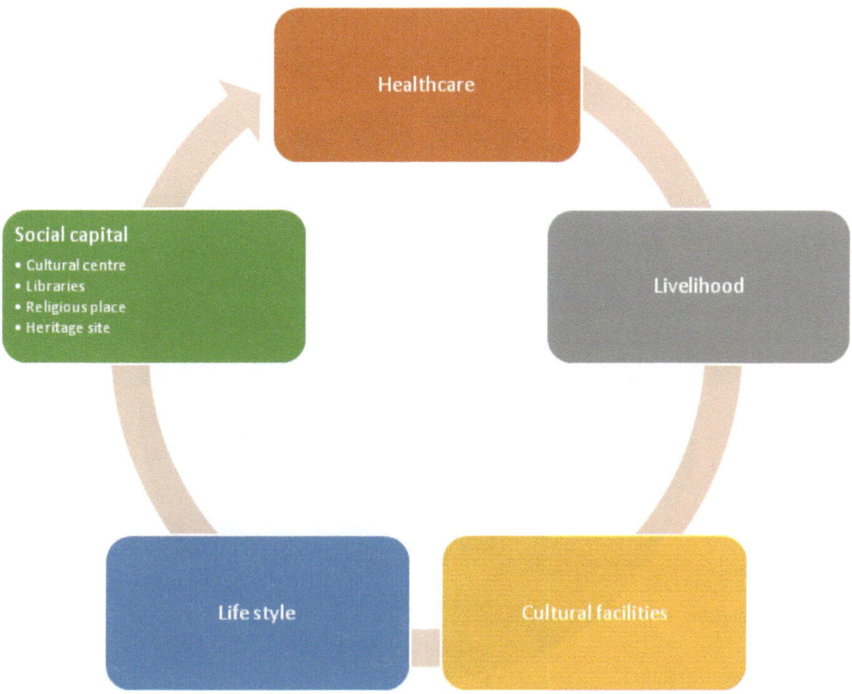

Fig. 2.5 Concept of smart living and quality of life

2.4.4 Smart Transportation

By creating an integrated ICT infrastructure to offer transportation and logistics services to businesses and individuals, intelligent mobility takes into account the creation of a secure, reliable, connected, and diverse transportation network while also cutting down on travel time and expenses as well as air pollution. Given that lowering air pollution and carbon emissions is the primary objective of smart transportation, clean transportation systems may include drones, hybrid cars, electric cars, and bicycles because they have the least polluting extents.

Smart transportation applies the smart traffic guidance framework, comprising software, signs, and human assistance, to assess the traffic scenario on the streets in response to the growth of urban life and the number of vehicles in metropolitan areas (Soomro et al. 2018). It has additionally established smart streets (Ahmed and Rani 2017) and intelligent parking lots (Ma and Xue 2020) in the city (Fig. 2.6).

Case Study: Online routing using open-source tool

Data is the most crucial component of network analysis in GIS applications. Depending on the data, the shortest path between two nodes may be substantially distinct. For instance, the user will receive inaccurate results from analysis if the road data lacks information regarding orientations (bidirectional or unidirectional). And

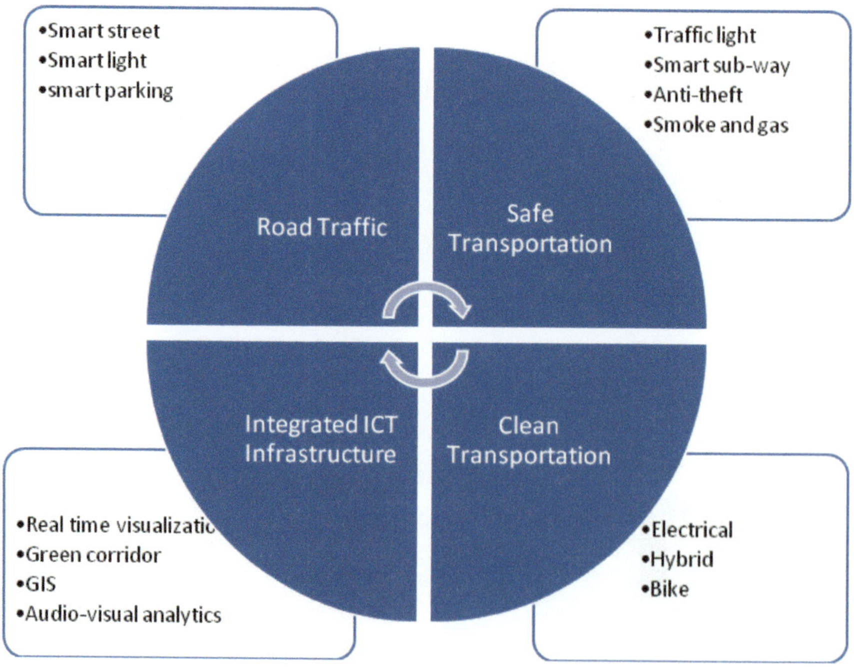

Fig. 2.6 Smart transportation and logistics services

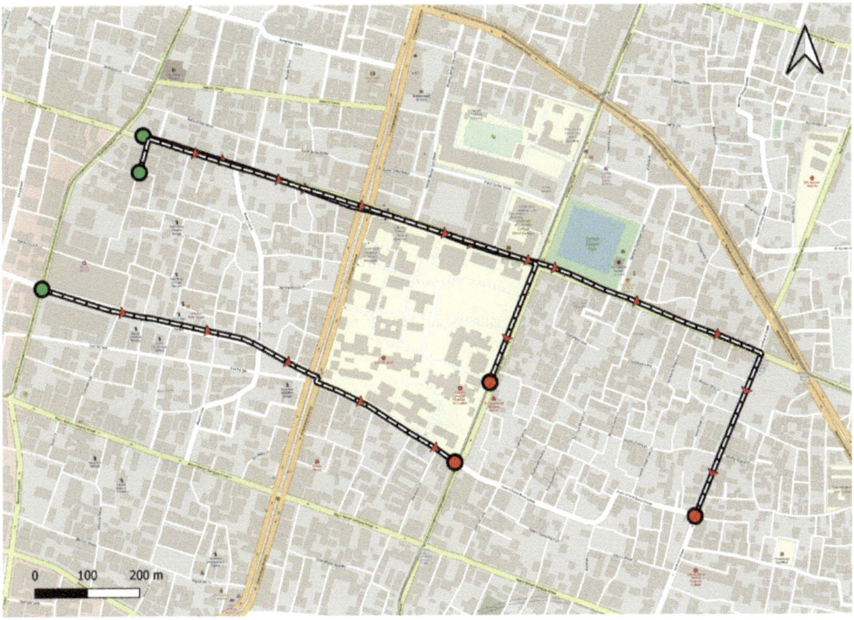

Fig. 2.7 Calculation of online routing based on QGIS open routing mapper plugin

if this is used in real life, it will be perilous. Today, using web maps (such as those from Google, Bing, Open Street Map, HERE, Yandex.Maps, etc.) for navigation is just as common as using them for browsing and exploring. Routes between cities and even between continents can be calculated by the user. Without having access to road network data, a user can compute a route in QGIS by pointing start and finish points using the Open Routing Mapper plugin (Fig. 2.7). With QGIS, results might be converted to several GIS formats, including ShapeFile, TAB, Spatialite, DXF, etc.

2.4.5 *Smart People*

Prior to infrastructure and technology, enhancing requires an intelligent, accountable, flexible, and productive citizen. None of the ready-made smart services will be used if there are no smart, demanding citizens. Consequently, rather than relying heavily on technology, the growth of smart city strategies accomplishes. Citizens have a significant role in this knowledge-based urban development since they not only consume information but also create it (Hernán et al. 2018). In order to raise the capabilities of the population of smart cities, it is vital to emphasize the significance of better education as well as the encouragement of creativity and independent thinking in people (Mohanty 2017) (Fig. 2.8).

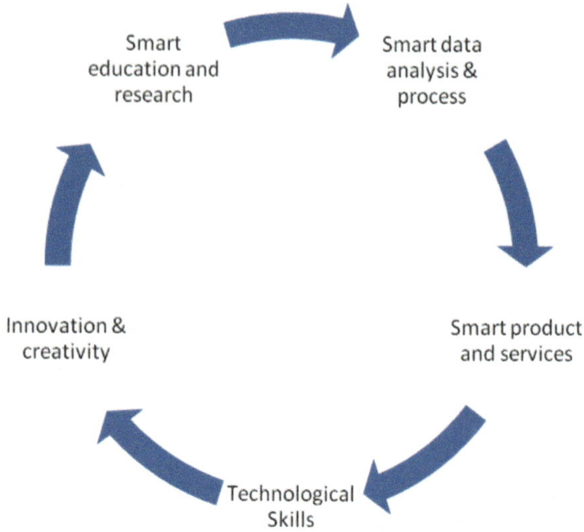

Fig. 2.8 Smart people used infrastructure and technology

One of the most crucial components of a smart city is its population. Smart people are those who possess technological know-how, inhabit a technologically advanced environment, have the ability to utilize virtual learning environments and learning spaces like smart education, and have succeeded in accomplished to foster greater creativity and independent thinking (Loo and Tang 2019). The focus is on human resources, capacity management, data processing, and assessment by the people themselves with the goal of executing decisions and providing goods and services in a context where clever people are present. The major objective of a city should be to increase its human capital, and variables for evaluating this include things like access to leisure activities, cultural diversity, and educational attainment.

According to Nam and Pardo (2011), the efficient manufacture of items like clothes, food, and tools may also serve to enhance the nation's economy and standard of living. Operators and IT firms are starting to track the incidence and identify additional high-risk regions as the importance of online shopping and food delivery startups in supplying residents' basic requirements has increased more than ever (Kylili et al. 2020). Digital technology should assist in tracking viruses to safely reopen urban infrastructures like airports, public transit, workplaces, and companies in an epidemic era—and this is not only restricted to today. Outbreaks should be prevented or kept to a minimum.

2.5 Urban Innovation and Sustainability

The term "smart city" is recent, and it can be thought of as a successor to the terms "knowledge city," "digital city," and "sustainable city" (Yigitcanlar 2006). Despite recent debates about the definition, there is no agreement about what constitutes a smart city (Hortz 2016). Considering the fact that a number of academics struggle with conceptualization, these concepts are not inconsistent, but rather partly overlapping (Cocchia 2014). As a whole, the word "smart city" is a broad term that encompasses a variety of sub-themes such as smart urbanism, smart economy, sustainable and smart climate, smart technology, smart energy, smart mobility, smart health, and so on (Cocchia 2014; Lara et al. 2016). Principles of sustainable urban growth and sustainable development give ideas of the production efficiency and use needed for residential, manufacturing, business, and recreational operation (Yigitcanlar and Kamruzzaman 2015). The Urban Innovation and Sustainability (UIS) seeks to strengthen the capacity of practitioners to cope with increasingly complex urban needs, obstacles, and opportunities.

According to Marsal-Llacuna et al. (2015), the preservation of the environment, the standard of life, and past accomplishments with livable and environmentally friendly towns should all be considered when evaluating a smart city, in addition to the arrangement of technological features. According to Lazaroiu and Roscia (2012), it should resemble a technological society that is interconnected, sustainable, convenient, pleasing, and secure. Smart cities use city data for traffic management, energy usage statistics, protection, and optimizing the efficiency of public services in order to understand how it functions in practice (Harrison et al. 2010).This new paradigm is stimulating the emergence of new smart city business niche suppliers that use technical tools to operate public services (Angelidou 2015).

Smart cities are divided into three dimensions, according to Nam and Pardo (2011): (a) technology (hardware and software infrastructures); (b) population (creativity, diversity, and education); and (c) institutions (governance and policy). As a result, technological innovations, population, and institutions aimed at achieving the idea of smart city stimulate sustainable growth and quality of life, while also encouraging prudent management of natural resources and enabling institutions to engage with innovation and improved services for residents, thus enhancing disputes and democratic engagement (Caragliu et al. 2011).

Taking all of these factors into account, Hiremath et al. (2013) describe sustainable urban development as bridging the gap between urban development and environmental conservation, with a focus on equality in wealth, housing, accommodation, basic services, social infrastructure, and transportation in urban areas. Many causes contribute to the increasing recognition of smart cities and associated principles, such as the fact that a significant portion of the world's population lives in cities, globalization, climate change, the dwindling supply of natural resources, and high demand. As a result, cities must provide residents with better and more customizable services (Angelidou 2015). A smart and sustainable city, as per Dhingra and Chattopadhyay (2016), has goals that must be met in an inventive, efficient, scalable, open, and resilient manner, such as:

- Progress the citizens' standard of living.
- Safeguard economic development by increasing job opportunities.
- Ensure residents' connections to social and community services in order to enhance their well-being.

- Establish a growth strategy that is environment friendly and sustainable.
- Make sure that essential infrastructure and amenities are provided in a timely and efficient way, including public transport, water supply and drainage, telecommunications, and other commodities.
- The goal of addressing environmental effects including changing the climate, in addition to.
- Provide a suitable regulatory and local governance structure to guarantee that policies are grounded in reality.

2.6 Urban Community, Resilience, and Quality of Life

Resilience has progressively been regarded as a concept that seeks to capture and coordinate a city's capacity to react, adapt, and cope with various changes, affected by climate fluctuations and changing at both a global and local level (Pike et al. 2010).With the advent of resilience thinking (Berkes and Ross 2013), the city's disaster response policy shifts from passive resistance and post-disaster mitigation to constructive adaptation, risk evaluation, and early warning systems (Ainuddin and Routray 2012). Cities are social–ecological structures that are known as combined, interconnected, and coevolving social systems and ecosystems (Folke 2006). Urban resilience is described as a comprehensive ability of an urban social–ecological system that is referred to by various interconnected subsystems in a city (Meerow and Newell 2015). Homeostasis, omnivory, high flux, flatness, buffering, and redundancy are six characteristics of a resilient framework provided by Wildavsky (1990). Multifunctionality, redundancy and modularization, (biological and social) diversity, multiscale networks and connectivity, and adaptive planning and design are among the five urban planning and design strategies advocated by Ahern et al. (2011) to achieve urban resilience. They also recognized eight key characteristics of buoyant cities, namely multiplicity, flexibility, modularity, creativity, close feedbacks, governance overlap, social capital, and ecosystem services.

The first component of the resilient neighbourhood framework is the external environment, which consists of two essential components of the adaptable community characteristics: the physical layout of the community's dwellings, entertainment centres, traffic roads, and landscape design; and the infrastructure for living, recreation, municipal services such as electricity and water, and municipal network technology. The hardware reassurance for managing various natural and man-made hazards, along with the physical strengthening of the robust population overall, comprises these two aspects. Public service, which encompasses information distribution, schooling and instruction, leisure activities, medical care, psychiatric therapy, and other forms of assistance provided to all members of the resilient society as a whole is the main component of the community endurance model.

The ***"spatial pattern of urban community"*** is the key to community structure's resilience of a changing climate, while measures like space distribution, composition, place, and function have immediate impact on structural resilience. Figure 2.9

illustrates how thoughtful development of community spatial patterns will increase communities' resistance to flooding brought on by climate change.

'Environmental elements of urban communities' are at the heart of a community's ability to adapt to changing and diverse needs. The types, amounts, and roles of the components are all indicators that determine their resilience. Rosenthal Park (Lower North Shore of Sydney, Australia) is an excellent example of a community's resilience in dealing with environmental and social issues by having environmental elements that are diverse in form and function (Fig. 2.10).

Because the "***service system***" in metropolitan areas is resident-oriented, it shows the durability of neighbourhoods during their operation. In the context of knowledge sharing and transfer between managers and citizens in urban communities, a strong connection network is created. Communities can deal with threats and adversities thanks to this network of relationships, as well as respond to urgent changes brought on by timely knowledge exchange. Service resilience is influenced by a number of factors, including the activities and individuals the services are intended for as well as the methods and instruments the services employ. Figure 2.11 depicts how the use

Fig. 2.9 The Skt. Kjeld community (located in Copenhagen, Denmark) represent the spatial pattern of urban community

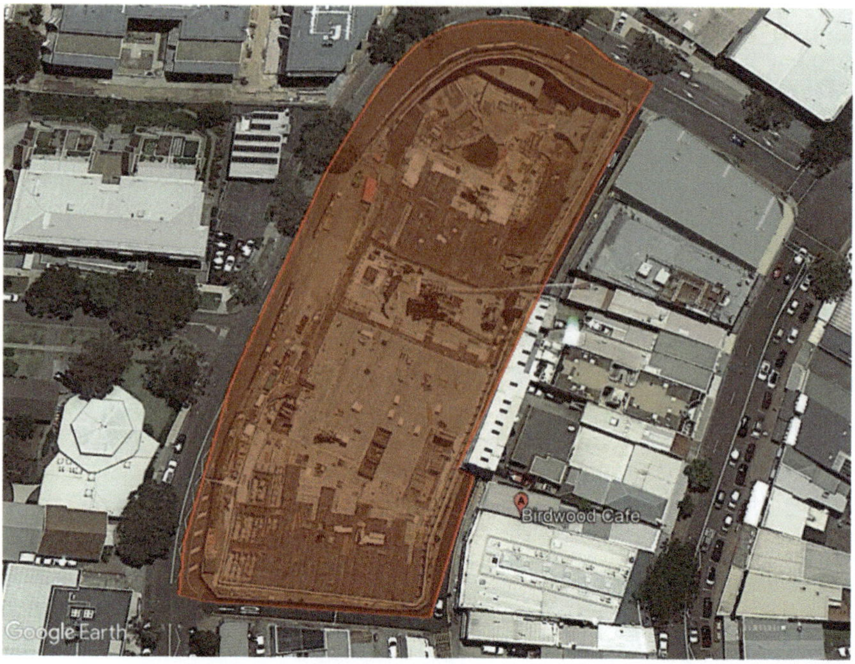

Fig. 2.10 Geographic location of Rosenthal park in Lane Cove community, lower north shore of Sydney, Australia

of scientific and humanistic know-how in the service system helps an urban society become more resilient to typhoons.

The "***management approach,***" which is organized by administrators in urban areas, is the way to put group resilience into practice at the organizational level. A resilient management structure enables the logical deployment and incorporation of community capital, as well as the optimization of community purposes and increased resilience in the face of threats and disasters. The methods and mechanisms of management, and also the managers' motivation and strategies, are all indicators that contribute to the management system's resilience.

2.7 Urban Informatics: Innovation and Challenges

Urban informatics is a new field that focuses on using new data sources to investigate urban systems (Thakuriah et al. 2016). The research of urban phenomena using a data science system of urban sensing, data mining and integration, modelling and analysis, and visualization to gain new insights that enhance computational science methods while also addressing domain-specific urban complexities is known as urban informatics (Kontokosta 2018). In order to apply new knowledge that

Fig. 2.11 Location of the New Jiangwan City in northeast of Shanghai

domain specialists may utilize to address problems, it focuses on applying urban computing and computer science tools to research, characterize, forecast, and, to a lesser extent, depict urban phenomenon. Urban informatics research and technologies have the greatest potential in four areas: (1) enhanced methods for complex urban resource management, (2) theoretical perspectives and information exploration of urban dynamics and processes, (3) strategies for urban interaction and public involvement, and (4) advances in urban management, planning, and policy analysis. With data-driven modelling techniques that are essentially theory agnostic, urban informatics creatively uses big data by updating or renovating pre-existing urban models and simulations based on a range of theoretical traditions, though these divergent research designs are starting to converge in some areas.

Urban informatics is the combination of computer science, physical science, decision scientific fields, operations assessment, business administration, and urban planning. As a result, the topic has been discussed in the literature across a variety of disciplines, with some researchers using a new language (or a lexicon borrowed from computing) to define conventional, quantitative social science approaches, while others use well-known computer science tools or physical models to analyse urban data (Zheng et al. 2014). A growing number of researchers are showing how big data can be used in urban settings (Kontokosta and Johnson 2017; Singleton et al. 2018). The purpose of this investigation is to enhance organizational decisions in city management by using new data sources and quantitative techniques from the computer and physical sciences (Hong et al. 2018; Kontokosta et al. 2018). Others point at how information and communication technologies (ICT) and the Internet of Things (IoT) are influencing the development of "smart cities" that monitor and automate several system-level functions that are currently done manually or in analogue (Zanella et al. 2014).

However, much of the literature lacks or undermines the long tradition of policy formation and philosophy that has attempted to figure out how to better incorporate emerging technology and quantitative approaches into decision-making processes (Batty 2014). The growing need for modern, integrated models that encompass behavioural and biophysical structures and mechanisms marks a significant divergence from domain-specific models based on generic hypotheses and rules of thumb (Alberti and Waddell 2000). Models developed from scarce input data, including those which depend solely on survey data, have had their practical utility limited in the past, leading to a justified backlash in policy making (Klosterman 1994). Correspondingly, the difficulty of integrating "proof" or "reality" into planning decisions has long been argued, as a reaction to the objective retrospective paradigm that arose in the mid-twentieth century and the notion of the planner as an apolitical technocrat (Healey 1992).

These complications can be divided into three categories: administrative, strategy and regulation, and planning and decision-making (Fig. 2.12). The effective management of systems and the equal distribution of public services are operational problems. Most applied data analytics research is focused on these, largely because they include problems with definable solutions, including mapping differences. These problems subject themselves well to data science techniques, such as machine learning, data mining, and information visualization. For example, situation awareness constraints are issues that call for a better comprehension of the location, function, or mobility of a certain resource or item. Targeting is a common requirement for agencies in order to optimize resource distribution. Cities must be able to allocate scarce resources efficiently while minimizing costs, whether it is police officers, building inspectors, or energy (Copeland 2015).

Recognizing causal inference is a skill that is frequently needed when working on policy matters that are centred on more significant social, economic, and environmental goals. A good example is program assessment, which involves estimating the impact of a planned or enacted policy. Different analytical tools are needed for policy difficulties than for operational ones. For example, more regression modelling and Bayesian inference—methods for separating the statistical or probabilistic impacts of specific variables—must be used. Furthermore, gathering fresh data that is specific to the problem at hand is frequently necessary to address policy-related concerns in an efficient manner. The emergence of big data and the potential to interpret large-scale data sets has ushered of a new era in public policy analysis (Foster et al. 2016). The capacity to evaluate policy choices in almost real-time and across several impact dimensions places additional demands on public decision-making frameworks that were established during an age when the evaluation and learning of interventions were restricted to survey data with minimal temporal and geographic precision.

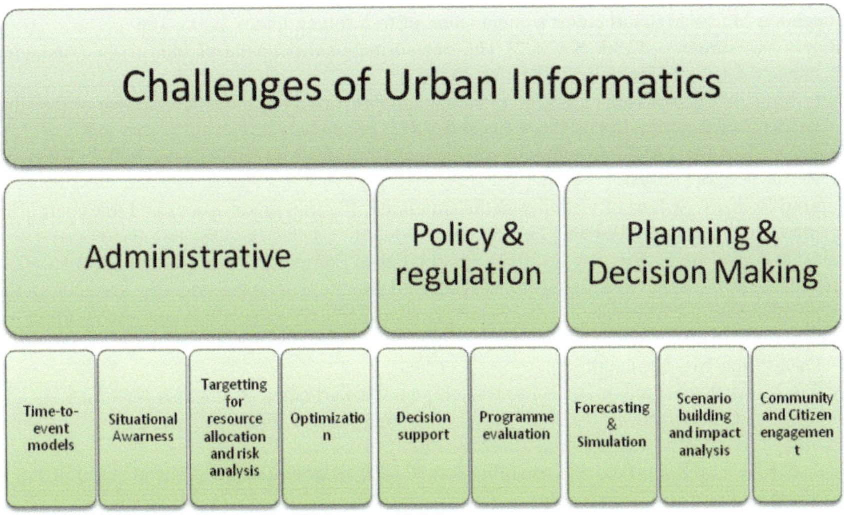

Fig. 2.12 Urban analytical challenges typology

Urban informatics can be used by planners to gain a greater awareness how urban space is used and to adjust information based on anticipated needs or projected demand. Data analytics and technology implementation are two distinct facets of planning issues. To begin, analytical applications include simulation, scenario modelling, and prediction, all of which can be used to evaluate the impact of various land use, transportation, as well as other transportation system decisions (Wang et al. 2017). Social media data that has been geotagged, for instance, may be utilized to produce time-dependent mobility patterns and high-resolution population predictions. This data can be used to forecast demand, plan traffic, and evaluate the effects of zoning and development projects.

References

Ahern J, Qin Y, Liu H (2011) From fail-safe to safe-to-fail: sustainability and resilience in the new urban world. Landsc Urban Plan 100(4):341–343

Ahmed S, Rani P (2017) Smart city initiatives in developing countries. Int J Adv Stud Sci Res 2(4):45–51

Ainuddin S, Routray JK (2012) Community resilience framework for an earthquake prone area in Baluchistan. Int J Disaster Risk Reduc 2:25–36

Alberti M, Waddell P (2000) An integrated urban development and ecological simulation model. Integr Assess 1(3):215–227

Al-Omari A, Alqaralleh B (2020) Smart waste management for sustainable urban development. Environ Res Technol 3(3):202–210

Alqahtani A, Kavakli-Thorne M, Kumar G (2020) Smart city framework and performance indicators. Sustain Cities Soc 59:102238

Angelidou M (2015) Smart cities: a conjuncture of four forces. Cities 47:95–106
Attaran M, Attaran S, Celik B (2022) The transformational potential of Industry 4.0 and smart cities. IEEE Eng Manag Rev 50(1):36–44
Batty M (2014) Can it happen again? Planning support, Lee's Requiem and the rise of the smart cities movement. Environ Plann B Plann des 41(3):388–391
Berkes F, Ross H (2013) Community resilience: toward an integrated approach. Society Nat Resources 26(1):5–20
Caprotti F (2018) Spaces of visibility in the smart city: flagship urban spaces and the smart urban imaginary. Urban Studies 56(12):2465–2479
Caragliu A, Del Bo C, Nijkamp P (2011) Smart cities in Europe. J Urban Technol 18(2):65–82
Cocchia A (2014) Smart and digital city: a systematic literature review. Smart city. Springer, Berlin, pp 13–43
Copeland (2015) Big Data in the Big Apple. http://www.spatialcomplexity.info/files/2015/06/Big-Data-in-the-Big-Apple.pdf
Deakin M, Al Waer H (eds) (2012) From intelligent to smart cities. Routledge, New York
Dhingra M, Chattopadhyay S (2016) Advancing smartness of traditional settlements – A smart village framework. J Clean Prod 124:278–290
Drohsel KM, Fey P, Höffken S, Landau S, Zeile P (2010) Mobile applications in urban planning, Weimar
General G, Martínez J, Tzatchkov V (2002) Urban infrastructure planning. J Infrastruct Syst 8(3):67–75
Giffinger R et al. (2007) Smart cities: ranking of European medium-sized cities. Vienna Univ of Technology
Foster I, Ghani R, Jarmin RS, Kreuter F, Lane J (eds) (2016) Big Data and social science: a practical guide to methods and tools. CRC Press, Washington, DC
Folke C (2006) Resilience: the emergence of a perspective for social–ecological systems analyses. Glob Environ Chang 16(3):253–267
Harrison C, Eckman B, Hamilton R, Hartswick P, Kalagnanam J, Paraszczak J, Williams P (2010) Foundations for smarter cities. IBM J Res Dev 54(4):1–16
Healey P (1992) Planning through debate: the communicative turn in planning theory. Town Plann Rev 63(2):143–162
Hernán M, Ortega M, Minguet P (2018) Citizen participation in smart city management. Sustain 10(3):939
Hiremath RB, Balachandra P, Kumar B, Bansode SS, Murali J (2013) Indicator-based urban sustainability – A review. Ener Sustain Dev 17(1):6–19
Hong B, Malik A, Lundquist J, Bellach I, Kontokosta C (2018) Applications of machine learning methods to predict readmission and length-of-stay for homeless families: the case of Win Shelters in New York City. J Technol Hum Serv 36(1):89–104
Hortz T (2016) The smart state test: a critical review of the smart state strategy 2005–2015's knowledge-based urban development. Int J Knowl Based Dev 7(1):75–101
Idwan S, Alsmadi I, Alshawabkeh M (2020) Smart cities – a survey. Int J Adv Comput Sci Appl 11(2):533–543
Innovation and Carbon (2016) Smart cities: smart vision for smart community. White Paper
Jong M, Joss S, Schraven D, Zhan C, Weijnen M (2015) Sustainable–smart–resilient–low carbon–eco–knowledge cities; making sense of a multitude of concepts promoting sustainable urbanization. J Clean Prod 109:25–38
Kontokosta CE (2018) Urban informatics in the science and practice of planning. J Plan Educ Res. https://doi.org/10.1177/0739456X18793716
Kontokosta CE, Johnson N (2017) Urban phenology: toward a real-time census of the city using Wi-Fi data. Comput Environ Urban Syst 64:144–153
Kontokosta CE, Boyeong H, Johnson NE, Daniel S (2018) Using machine learning and small area estimation to predict building-level municipal solid waste generation in cities. Comput Environ Urban Syst 70:151–162

Klosterman RE (1994) Arguments for and against planning. Town Plan Rev 65(1):1–12
Kumar T, Dahiya B (2017) Smart cities development in India: an appraisal. J Urban Manag 6(2):24–34
Kurniawati W (2018) Urban equity in public space. A comparison research between traditional and modern public space in Indonesia. Dissertation, ULB TU Darmstadt
Kurniawati Y, Prihantini D (2018) Smart city planning for sustainable urban development. Int J Adm Sci Organ 25(3):105–112
Kylili A, Fokaides PA, Jimenez P (2020) Smart cities and circular economy. Renew Sustain Ener Rev 134:110852
Lara A, Costa E, Furlani T, Yigitcanlar T (2016) Smartness that matters: comprehensive and human-centred characterisation of smart cities. J Open Innov 2(8):1–13
Lazaroiu GC, Roscia M (2012) Definition methodology for the smart cities model. Energy 47(1):326–332
Loo BP, Tang T (2019) Transport planning and smart mobility. J Transp Geogr 79:102481
Low S, Taplin D, Scheld S (2005) Rethinking urban parks: public space and cultural diversity. University of Texas Press, Austin
Marsal-Llacuna ML, Colomer-Llinàs J, Meléndez-Frigola J (2015) Lessons in urban monitoring taken from sustainable and livable cities to better address the smart cities initiative. Technol Forecast Soc Chang 90:611–622
Ma L, Xue F (2020) Big data analytics for smart city. Sustain Cities Soc 54:101879
Meerow S, Newell JP (2015) Resilience and complexity: a bibliometric review and prospects for industrial ecology. J Ind Ecol 19(2):236–251
Mohanty S (2017) Smart city: concept and challenges in India. Int J Res IT Manag 7(4):1–15
Moore T, Van Pham D (2019) Smart city indicators and governance frameworks. J Urban Manag 8(2):27–36
Mumford L (1966) The city in history: its origin, its transformations, and its prospects. Penguin
Nam T, Pardo TA (2011) Conceptualizing smart city with dimensions of technology, people, and institutions. In Proc 12th Annu Int Digit Gov Res Conf (dg.o '11) pp 282–291
Newman P, Kenworthy J (1999) Sustainability and cities: overcoming automobile dependence, Washington, DC: Island Press
Nin J, Villatoro D, Cámara D (2019) Crowdsensing and smart city planning. Sensors 19(8):1843
Nitti M, Pilloni V, Giusto D, Popescu V (2017) IoT architecture for smart cities. IEEE Int Thin J 4(5):1224–1235
Pike A, Dawley S, Tomaney J (2010) Resilience, adaptation and adaptability. Cambridge J Regions Econ Society 3(1):59–70
Radwan AH, Morsy AAG (2018) Smart urban public spaces-towards a better city life (May 27, 2018). Forthcoming, Proceedings of CPUD'18 conference, organized by DAKAM (Eastern Mediterranean Academic Research Center), Available at SSRN: https://ssrn.com/abstract=3277807
Ravazzoli E, Torricelli GP (2017) Urban mobility and public space. A challenge for the sustainable liveable city of the future. J Public Space 2(2):37–50. https://doi.org/10.5204/jps.v2i2.91
Roy M, Howlett M, Karpatne A (2018) Urban resilience and the smart city agenda. J Urban Technol 25(4):3–32
Singleton A, Spielman SE, Folch DC (2018) Urban analytics. SAGE Publications Ltd., Thousand Oaks, CA
Soomro K, Mirza A, Abro A (2018) Role of ICT in smart city development. Int J Adv Comput Sci Appl 9(2):80–88
Thakuriah P, Tilahun N, Zellner M (2016) Big data and urban Informatics: innovations and challenges to urban planning and knowledge discovery. In: Thakuriah P, Tilahun N, Zellner M (eds) Seeing cities through Big Data: research methods and applications in urban informatics. New York, Springer, pp 11–45. ISBN 9783319409009. https://doi.org/10.1007/978-3-319-40902-3_2

Townsend AM (2013) Smart cities: Big data, civic hackers, and the quest for a new utopia. WW Norton & Company, New York

Villa-Arrieta JA, Sumper A (2019) Smart city concepts: a review. Sustain Cities Soc 45:43–54

Wang L, Qian C, Kats P, Kontokosta C, Sobolevsky S (2017) Structure of 311 service requests as a signature of urban location. PLoS ONE 12(10):e0186314

Wildavsky A (1990) Public policy: the art of the possible. Boston, MA: Little, Brown and Company

Yigitcanlar T (2006) Australian local governments' practice and prospects with online planning. URISA J 18(2):7–17

Yigitcanlar T, Lee SH (2014) Korean ubiquitous-eco-city: a smart-sustainable urban form or a branding hoax? Technol Forecast Soc Chang 89:100–114

Yigitcanlar T, Kamruzzaman M (2015) Planning, development and management of sustainable cities: a commentary from the guest editors. Sustainability 7(11):14677–14688

Zanella A, Bui N, Castellani A, Vangelista L, Zorzi M (2014) Internet of Things for smart cities. IEEE Internet Things J 1(1):22–32

Zheng Y, Capra L, Wolfson O, Yang H (2014) Urban computing: concepts, methodologies, and applications. ACM Trans Intell Syst Technol (TIST) 5(3):38

Zygiaris S (2013) Smart city reference model: assisting planners to conceptualize the building of smart city innovation ecosystems. J Knowl Econ 4:217–231

Chapter 3
Services, Decentralized Planning, and Disparity

3.1 Introduction

Governments in developing countries have crucial issues in delivering basic amenities to their population, such as potable water, sanitation, suitable housing, and electricity, as the world's fastest urbanizing region. Although local governments are generally responsible for providing these services, few developing countries are fiscally decentralized, and municipal governments lack the financial capacity to adequately implement these services. Because of their small revenue bases, sub-national governments are frequently compelled to rely on intergovernmental transfers from the federal government. Because of the stress that cities in developing nations are experiencing as a result of demographic change and economic liberalization, a slew of concerns relating to efficient city government have arisen. Unfortunately, due to a shortage of resources or inappropriate power decentralization, most city governments are ill-equipped to address some of these issues. As a result, citizens have been provided with insufficient and ineffective public goods and services. Partnerships between municipal governments, civic society, and the commercial sector become critical in this scenario to drive and establish responsive governance systems.

It is critical to recognize that, in order to effectively govern and plan metropolitan regions, local city governments must be given the authority to carry out all administrative functions. This type of devolution of power and funding has not occurred in most developing countries, and as a result, local governments are unable to function or perform efficiently. Figure 3.1 shows the indicators of good governance. It assures that societal priorities in terms of politics, social issues, and economics are founded on widespread consensus. It guarantees that decision-making takes into account the viewpoints of minorities, lessens corruption, and gives voice to the most marginalized members of society.

G. S. Bhunia and P. K. Shit, *Urban Morphology and Sustainable Smart Cities*,
https://doi.org/10.1007/978-3-031-97473-1_3

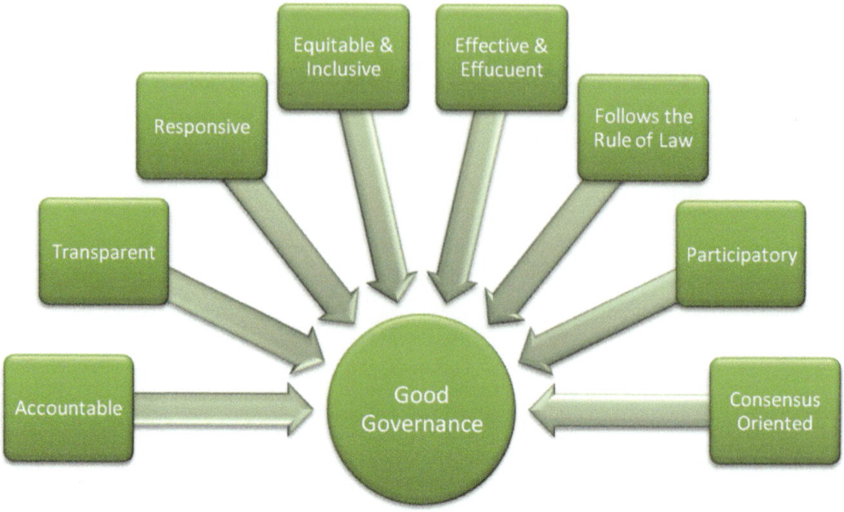

Fig. 3.1 Indicators of good governance

3.2 Cities Safety, Security, and Smart Public Services

The concept of safe city enhances the efficacy of procedures in the field of urban safety and health by lowering crime and terrorist threats, enabling residents to live in an environmentally friendly and healthy environment with easy utilization of high-quality health care. This is achieved by integrating with the idea and utilizing organizational, hardware, software, and physical means of safeguarding the environment. Globally, cities are facing an increasing public security concern. Globally, cities are overloaded, public resources are few, and criminal enforcement is more challenging than ever. Most video surveillance systems are no longer functional. Police investigations are frequently hampered by video network blind spots and low-quality imagery. In the realms of traffic and data management, large cities create increasingly complex difficulties. At the same time, these data present great opportunity to increase sustainability and make cities smarter.

Cities serve as the focal point of the country's political, economic, and cultural activity. They also house the bulk of crime and other hazards, such as industrial incidents, catastrophic events, human trafficking, and terrorist activity. Ensuring public safety is a top priority for city governments, as it helps them reach population growth targets while also improving the quality of life and drawing in visitors, companies, and skilled labour.

Fig. 3.2 Smart solutions can improve multiple aspects of the quality of life in cities

Data storage and retrieval issues mean incident data can take a long time to reach command staff, and data loss can completely derail investigations. New technology, on the other hand, is beginning to become accessible to assist police in combating crime more efficiently. Huawei, a prominent worldwide supplier of information and communications technology (ICT) approaches, is a trailblazer in the domain of "smart city" solutions. These solutions utilize a linked network of devices, software, and cloud storage systems to facilitate more efficient collaboration between both the public and private sectors (Fig. 3.2).

Although technology cannot provide a quick answer for crime, administrators can utilize data to better allocate precious resources and staff. Prescriptive analytics, on the other hand, takes a step further by predicting crime to stop incidents before they happen. For instance, real-time crime visualization employs statistical evaluation to highlight trends. Technologies such as smart surveillance, home security systems, and gunfire identification can speed up law enforcement's response time in the case of an incident. But the application of data-driven policing needs to be done so in a way that upholds civil rights and forbids criminalizing particular populations or demographic groups.

3.2.1 *Smart Transport System*

The smart transport system is one of the most talked-about systems. It impacts all city residents along with administrative and commercial entities due to its many effects, which include pollution, traffic jams, and travel times around the city. The fact that inadequate transit infrastructure may kill a city's vitality and economy and cause it to lose valuable real estate is another indication of its importance. Therefore, housing and land use policies must be tied to transportation issues. The population density of the city must also be taken into account. The densest areas are becoming the hubs, where it is most convenient to gather shops and services, so that no one needs to walk—or drive—long distances to get what they require (The Worldwatch Institute

Island Press 2016). As part of the smart transportation framework, the smart transport layer lowers travel times, noise, costs, unneeded economic charges, and congested roadways to lower air pollution and facilitate simplicity of access. By incorporating innovative technology into establishing and development, it additionally enhances the effectiveness of transportation and movement of people and goods in the city and its surrounding region.

Through the implementation of technology, planning, development, traffic laws, and supervision, the safe layer of the smart transportation system lowers the frequency of various traffic accidents, enhancing the security of both the integrity and security of traffic infrastructure in addition to the protection for every user of the system and alternative forms of transportation. The following features should be included in the smart transport system's layers:

- *Functional layer*: Roads and other transportation networks of high quality, bridges and tunnels, modes of transportation, road signs, and pertinent data.
- *Safe layer*: Rules governing traffic, secure parking lots, sources of data processing from essential infrastructure elements (supervision management), tools for forecasting the impact of multiple crisis scenarios on traffic, and the planning of backup and evacuation pathways.
- *Smart layer*: UMT embraces traffic with efficient back-to-back interconnections and the flexibility to modify services to accommodate evolving commuter demands; it also favours and supports the use of bicycles, pedestrians, and emission-free cars; intelligent collecting tools; intelligent parking technologies; real-time informative tools for drivers; efficient traffic signal and signalling systems with real-time modifications.
- *Design other features*: Because of the vast and continually expanding pool of options, discovered features are not definitive, though they can be further separated into more specialized domains, this article is not meant to go too far into any particular structure.

Case Study: Find Direction of Road Network

The task of determining the shortest path between two sites/locations is one of the most frequent spatial problems we encounter on a daily basis. By computing the total cost between two places (the source and destination sites) in a city, a GIS tool may be used to determine the shortest travel length of time. System calculates the path as well as cost, optimizes by length and exports path to a vector layer. The road network data used in this analysis is obtained from Jhansi city of Uttar Pradesh in India and the distance spacing is calculated based on the value of latitude and longitude obtained from the Open Street Map API. Using a numerical network property, such as form length, the shortest path is determined. A shortest path trace can be used to create paths that are cost- or distance-based. For this analysis, the node combination algorithm is taken into account, which essentially merges the two nodes with the smallest distance (Fig. 3.3).

Fig. 3.3 Finding direction of road network

3.2.2 *Smart Government System*

By integrating smart information systems and communications infrastructure, the smart layer of the smart government system generates transparent governance and effective digital government, and by implementing innovative policies, it combats the city's financial, environmental, and service obstacles in order to achieve long-term advantages and smart innovation (Mutiara et al., 2018). The use of software, construction, organizational resolutions, and paraphernalia to create a safe government layer improves the cybernetic security of used processes and connections, as well as the cybernetic security of government employees, objects, and amenities, as well as the cybernetic security of people and their belongings in an administered region, in order to ensure the adaptability of government functioning and to boost law compliance and the telecommunications security.

Digital government should digitize municipal services and improve collaboration both within government and between government and citizens, allowing both sides to better comprehend one other's reasons and requirements and allowing citizens to participate in city planning (Rochet 2015). Smart government relies on transparent governance, as well as the quality and accessibility of information and services. The following characteristics of the smart government system:

- *Functional layers*: The law, the executive branch's hierarchical and functional structure, government items, infrastructure, and offices, and pertinent information.
- *Safe layer*: Law enforcement, legal system, and secure Internet connections and communication technologies (including Intranet).
- *Smart layer*: Smart government initiatives, user-friendly and effective digital administration, efficient communication platforms, and open data for the general public are all examples of smart development policies.

Case Study: Intelligent Property Tax Management System

The ability of local authorities to operate properly, honestly, and with platforms to engage with residents is one of the most important functions of the government. Greater accountability and multiplied income increase for local authorities are two benefits of GIS-based property and tax management. Urban Local Bodies (ULBs) can correctly register properties in accordance with the property tax jurisdictions of commercial, residential, or industrial thanks to GIS-based systems, which also give a 360-degree perspective of the taxable income stream.

Methods:

Field surveys employing the property tax form were used to gather the primary data, or categorize property data. Wards boundaries and tax zone boundaries are examples of secondary data that have been gathered from the relevant departments. It was suggested that an identification known as BUID serve as the special link between the building's spatial data and the property survey form. The geocode tool has been used to map the survey data on the QGIS platform. The three associated layers—building, properties, and unit—were developed as a result of the survey data. Since "property" is the primary criteria for property taxation, it can be a single building, a portion of a building, or a group of buildings. To enable the mapping of real-world attributes using polygons, every building has been given a Building Unique Identification Number (BUID)throughout the digitization process (Fig. 3.4).

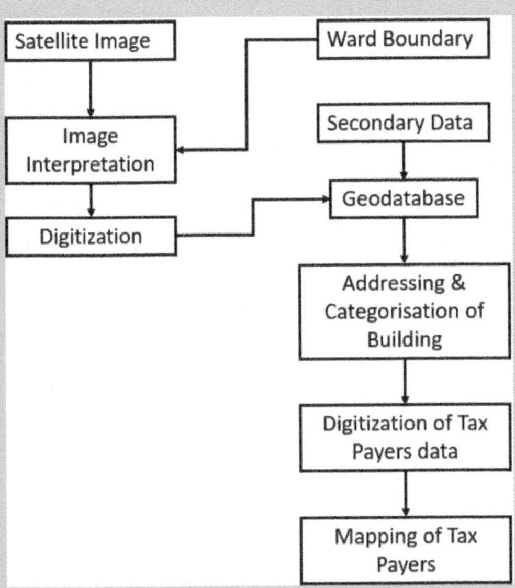

Fig. 3.4 Process of tax management system

Results:

Any management and planning operations for any place must be based on an understanding of the local reality in terms of features, structures, and economic activity. This necessitates digitizing all of the existing physical elements, including any buildings, lands, and roads, that are located inside the municipal boundaries. GIS helps in tracking of the revenue from property taxes by creating a platform that geographically relates all property-related data, including the number of floors in each building, the total built up area of each building, individual plot areas, locality details, and municipal colony boundaries, to the tax calculation principles (Fig. 3.5).

Fig. 3.5 Monitoring and property taxes from different sites

Benefits:

- Monitoring the payment of property taxes on a regular basis.
- An estimate of the revenue received from property taxes
- Zone/ward-wise tax collection system.

3.2.3 Smart Citizen and Smart Education System

The smart layer of the smart citizen and smart education system depicts collaboration and interactions between the government, smart city participants, teachers, citizens, and students, with teachers guiding students through all-round theoretical and practical aspects in a friendly environment (Cardullo and Kitchin 2017; Bull 2016). The majority of the smart education system and the protection of smart citizens are under the purview of the safe city concept. The safe layer of the smart citizen and smart education system contributes to enhanced safety by fostering the development of morally sound citizens who are knowledgeable about crisis behaviour, first aid, and survival essentials. In certain cases, it also raises professionals with advanced degrees in safety-related fields of study. The smart citizen and smart education systems have the following features.

- *Functional layer*: Factors to take into account include students, instructors, citizens, schools, expertise and pertinent data, and culture.
- *Smart layer*: Government, e-Government, smart city stakeholders, smart apps, technology, and gadgets for regular citizen usage, as well as businesses working with students.
- *Safe layer*: Morality, first aid, crisis behaviour, survival essentials, physical education, safety-oriented higher education, the ability to acquire self-defence, and helpdesk and psychologist services are all examples of educational themes.
- *Design and other features*

3.2.4 Make Daily Commutes Faster and Less Frustrating

Every working day, tens of millions of people in cities throughout the globe spend their mornings and evenings stuck in traffic or crammed into congested buses and trains. To raise quality of life, the daily commute has to be enhanced.

By 2025, cities that use smart mobility technologies might see an average reduction in commute times of 15% to 20%, with certain individuals potentially benefiting even more. The prospective benefit of each application varies greatly depending on the city's population, the state of the transit system, and commuter patterns. In a densely populated city with extensive transit, smart technology may save the typical passenger almost fifteen minutes each day. The reduction in commute time might be as little as 20 to 30 min a day in a growing city where commutes are lengthier.

Apps that lessen traffic congestion work well in cities where driving is prevalent or when buses are the primary mode of transportation. Intelligent traffic signal synchronization has the potential to reduce average commute times by more than 5% in growing cities where buses are the primary mode of transportation. Real-time navigation directs cars to the fastest route and alerts them about approaching delays. By directing users to open spots, smart parking applications help people avoid wasting time driving around town in circles.

3.3 Cities Well-Being, Health, Utility Services

Many of the world's fastest-growing cities are grappling with the same health issues that plagued industrialized countries in the nineteenth century. Because so many elements influence personal health and well-being, pinpointing the possible impact of the physical urban environment on health is difficult. The majority of individuals still think that genetics and medical treatment are the main variables that determine health, but we now know that factors related to society and the environment also play a big role. In 2012, factors related to living conditions, transport, polluting the air and noise, and other aspects of the external environment accounted for around 23% of all fatalities globally. Built environment professionals may contribute to the minimization of negative health impacts and the maximization of positive health benefits via the design and development of facilities and communities. Concurrently, cities in developed nations are dealing with new problems, partly brought on by the layout of our cities, the transit systems that support them, and the manner of life we have adopted. Property experts have been at the centre of decision-making regarding the physical environment for more than a century and a half, whether at the scale of land development, preferences about the sorts of property to be created, or the cost at which expansion is to be accomplished. When applied on the fringes of rising cities, the benefits of enhanced development standards can be felt quickly through the elimination of diseases like cholera and typhoid. In industrialized economies, the benefits are more long-term, as people live longer and healthier lives. In a time when financial resources are constrained, it is simple to see how arguments are formed to justify why costs that have no evident immediate return cannot be committed.

The term "social determinants of health" refers to a group of features that influence our health outside of our genes. These are "the circumstances in which people are born, grow up, live, work, and age, as well as the systems put in place to deal with illness," according to the World Health Organization (WHO). The physical buildings and urban settings in which we spend a great deal of time can have a significant effect on our health and well-being. Through policies spanning several departments, city officials are increasingly affecting health and well-being. It addresses both planning policies and related activities across city departments with the goal of improving inhabitants' health and well-being as the city expands. The desire to alleviate health inequality is a significant component of the plan. These challenges are compounded in low-income countries, where the most impoverished residents lack access to basic water and sanitation facilities and buildings are rarely constructed to code, creating significant dangers, particularly during natural disasters.

3.3.1 Smart Infrastructure and Smart Energy System

The smart environment system and the smart layer of the smart infrastructure and energy system work together to improve the city's grey infrastructure, make buildings more interactive, energy-efficient, digital, environmentally friendly, and comfortable, supply a smart grid, dependable telecommunications infrastructure, adaptable and all-encompassing energetic services, and make it possible to use intelligent energy systems and sources of renewable energy. Infrastructure safety in the city is again the responsibility of the safe city concept. The protected layer of smart infrastructure and smart energy systems contributes to infrastructure safety by following construction codes, interconnecting infrastructure with sensors that may notify competent workers about dangers, and offering backup energy systems. Smart infrastructure and smart energy system features:

- *Functional layer*: Buildings, electricity, HVAC systems, and telecommunication infrastructure are examples of grey infrastructure and pertinent information.
- *Smart layer*: Examples of intelligent energy systems include smart grids, energy-efficient power plants, energy-saving and control programmes, smart home systems, and waste energy rehabilitation programmes.
- *Safe layer*: Building codes, sensors, infrastructure element supervision, preventive measures for key infrastructure and energy systems, and backup energy systems are just a few of the tools available.
- *Design and other features*

3.3.2 Social Connectedness

If mechanisms are put in place to facilitate communication between the general public and municipal authorities, city governments could respond more quickly. A large number of local departments use social media, and some have even developed their own interactive citizen apps. In addition to giving information, these channels give locals a way to gather data, raise difficulties, and offer their opinions on planning matters. With Paris's introduction of a collaborative budget, anybody may submit project suggestions and then cast an Internet-based vote to choose which should be financed.

3.3.3 Smart Safety and Smart Healthcare System

We provide a unique, cross-sectional system smart safety and smart health care, similar to the previous one. Because every system must be secure, this system incorporates all other smart city and safe city concepts. It also integrates the concepts of safe cities and smart cities. It suggests that this system is just partially encompassed by both categories. By lowering crime and terrorist threats and enabling citizens to prosper in an environmentally friendly and safe environment with effortless possession of high-quality health care, the smart safety and smart healthcare system evaluates the effectiveness of procedures throughout the field of urban safety and health. It does this by utilizing smart technology, the natural environment, physical, organizational, hardware, and software means of safeguarding.

The urban environment is made up of a variety of interconnected components that have direct and indirect effects on one's health and well-being. The city may have an impact on people's health on many levels, from macrosystems like air pollution to microsystems like providing benches and street lights to encourage people to stroll around their neighbourhood. Figure 3.6 shows a variety of physical aspects of metropolitan environments that affect people's happiness and health. Managing aspects of the urban environment that have an impact on health is the duty of multiple ministries within municipal government. Outside of city hall, private sector public organizations will administer services, buildings, and infrastructure. To increase the city's health impact, stakeholders must collaborate and work independently.

Because of the system's unique, cross-sectional nature, its features are not segregated into smart and safe layers, implying that the entire system belongs to both

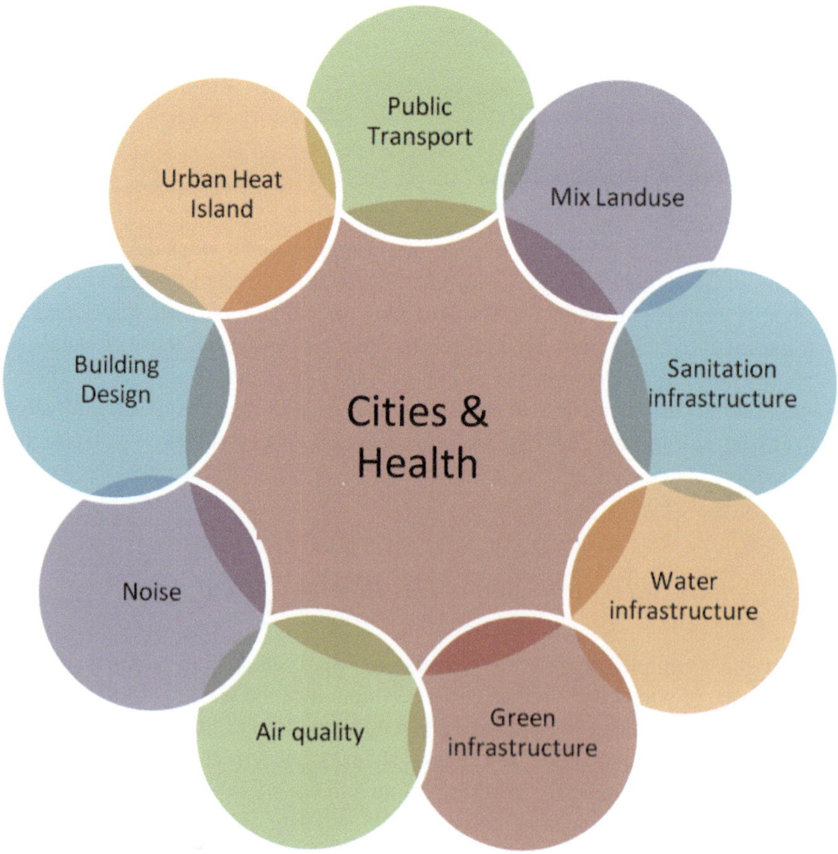

Fig. 3.6 Urban environmental factors

the smart and safe layers at the same time. As a result, this system consists of only three layers: the functional layer, the core layer (smart and safe character), and the aesthetic layer. The possibilities for improving safety vary depending on where you are in the disaster management cycle. There are four phases in this cycle, divided into two periods: pre-crisis and post-crisis. The first is a period of preparation, while the second is a period of reaction.

To get a thorough understanding of a smart city system, it is necessary to go through each phase and search for future possibilities. Optimization of required employees, time, equipment, resources, information sources, and information flow should be the main areas of improvement. We must look for prospective improvements in each phase by asking ourselves the following questions.

- What benefits might a smart component of this system provide in terms of reducing the quantity of employees and time required to handle safety-related duties during this stage of the disaster management cycle?
- What benefits can a smart element of the model provide in terms of reducing the number of pieces of equipment required to perform safety-related functions, and what smart equipment can be employed at this stage of the disaster management cycle to increase safety?
- What benefits can a smart component of this system provide in terms of reducing costs for equipment and labour required to handle safety-related tasks, minimizing financial and material losses, and identifying additional financial and informational resources that will be useful during this stage of the disaster management cycle?

3.4 Urban Ecosystem Services and Sustainable City Planning

Environmental pressures increase as urbanization, industrialization, and consumption increase. Building automation technologies, dynamic electricity pricing, and some mobility applications could all work together to reduce emissions by 10 to 15%. When it refers to the smart city concept, most people think of green fields and parks in cities. They are critical, yet they are not the only component of the system. The advent of smart technologies is set to fill the world with technology that will enable us efficiently engage with the environment, both green and grey (Nosek et al., 2017). The smart layer of the smart environment system, which uses smart devices such as sensors and communicational technologies, as well as green infrastructure, creates an entertaining, interactive, automatic, data-generating, and assessing environment with sufficient leisure public spaces in addition to making the environment itself support decision-making, to enable effective control and use of the environment, and suitable landscape (Holubčík et al., 2012).

In cities with high household water usage, water consumption tracking, which combines improved metering with digital feedback messages, can drive people towards conservation and lower consumption by 15%. Seepage from pipes is the most common form of water waste in many parts of the developing world. The use of sensors and analytics can reduce the risk by up to 25%. Pay-as-you-throw digital tracking systems, for example, can cut solid waste per capita by 10 to 20%. Overall, communities can save 25 to 80 L of water per person per day and 30 to 130 kg of unrecycled solid waste per person per year.

Air quality sensors cannot immediately solve pollution's causes, but they can identify the origins and serve as a starting point for further action. By regularly analysing pollution sources and controlling traffic and development accordingly, Beijing was able to lower harmful airborne contaminants by around 20% in less than a year. Individuals can take protective precautions by sharing real-time air quality information with the public via smartphone apps. Depending on existing pollution levels, this could lessen harmful health consequences by 3 to 15%.

In a broader, regional context, safe layer is concerned with animal welfare and land use, as well as the preservation of ecological diversity and the food safety of rural agricultural product. The smart environment system has the following functionalities:

- *Functional Layer*: Animals, pertinent information, land and agricultural production management, green infrastructure, blue infrastructure (water and wastewater management), grey infrastructure (accessed with smart infrastructure system), community (shared with smart citizen system).
- *Smart Layer*: Smart devices, smart connections, environmental control centres (flats, buildings, city regions, etc.), recreational playfields, and landscaping.
- *Safe Layer*: Sanitation features in environment, rainwater management, heat and emission control, decision-making support programmes, health monitoring, parks, and food safety governance are all examples of natural sanitation features.
- *Design and Other features.*

3.5 Urban Migration and Regional Disparity

A large proportion of people on the planet reside in cities, and this number is rising. The proportion of the world's population living in cities is anticipated to rise from 55% in 2018 to 60% in 2030 (UN 2018). Most of the fastest-growing cities in the entire globe are located in Asia and Africa. Between 2018 and 2050, Africa's urban population is expected to triple, while Asia's is expected to grow by 61%, resulting in Asia (52%) and Africa (21%) hosting the majority of the world's urban population by 2050 (UN 2018). Urbanization has traditionally been fuelled by internal and overseas migration, which offers opportunities and challenges to governments, migrants, and municipalities alike. Municipalities are starting to integrate migration into their urban planning and implementation as they become more and more seen as important participants in migration management (Fig. 3.7).

Beijing, Berlin, Brussels, Buenos Aires, Chicago, Hong Kong SAR, China, London, Los Angeles, Madrid, Moscow, New York, Paris, Seoul, Shanghai, Singapore, Sydney, Tokyo, Toronto, Vienna, and Washington DC are anticipated to house

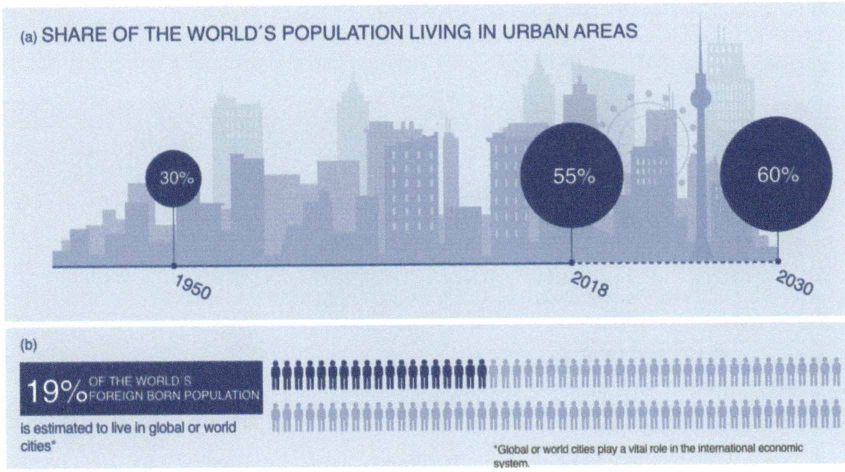

Fig. 3.7 Urbanization and migration (*Source* UN DESA, 2018, IOM)

one out of every five foreign migrants (International Organization for Migration (IOM) 2015). International migrants made up around 20% of the overall population in 18 of these cities. In some cities, the proportion of foreign-born people in the total population far outnumbers the global average (about 3.5%). Various kinds of migration, to varying degrees, contribute to urban growth and variety. International migration is one of the main drivers of population variety in industrialized nations, but domestic migration is most common in underdeveloped countries (IOM 2015), in addition to demographic expansion as births outnumber deaths. In other countries, such as China and Thailand (80%), Rwanda (79%), Indonesia (68%), and Namibia (59%), rural-to-urban migration and reclassification of what is considered urban jointly accounted for more than half of urban growth (UN, 2018). Circular and transitory migration can be observed in many Asian and African cities, particularly in China and India, as well as Ghana and Kenya (Fig. 3.8).

Mixed migration is a newish word that attempts to represent the intertwined and diverse reasons of all people's mobility, regardless of status. "Cross-border movements of people, including refugees escaping persecution and conflict, victims of trafficking, and those seeking better lives and opportunities," is what mixed migration refers to. People involved in mixed migration are motivated by a number of motivations, and they have varying legal positions and are exposed to a variety of risks. While crossing national borders is sometimes categorized as either "forced/involuntary" or "voluntary," the reality is far more complicated and nuanced (Fargues 2020). For travellers who may not be capable of obtaining permanent residency or who may not have glanced themselves in for reasons specified in the 1951 Refugee Convention or regional refugee instruments, but who might have thought they were permitted to flee for many different kinds of interconnected variables including economic, political, social, religious, or ethnic causes, a mixed migration sensor contributes to broadening

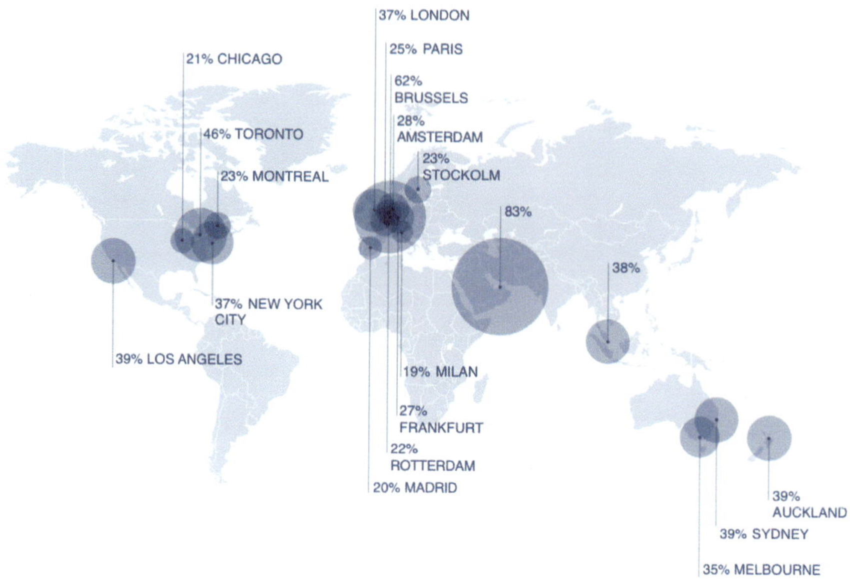

Fig. 3.8 Foreign born population in some major world cities (*Source* IOM, GMDAC 2019, www.migrationdataportal.org)

the safety space. These folks frequently encounter the same dangers, have comparable requirements, and travel along the same pathways. They may, meanwhile, fall through the protective net, rights protection, and aid because contemporary geopolitical guidelines only support two concepts: migrant and refugee (IOM Displacement Tracking Matrix 2017). Given the concealed, cross-border, and transitory character of mixed migration, there are just a few large-scale quantitative datasets on the subject to date, making reliable data collection extremely difficult. Furthermore, different definitions of mixed migration have an impact on data collecting and comparability.

Whether on the ground or on the final path, the International Organisation for Migrants' (IOM) Displacement Tracking Matric (DTM) seeks to quantify and characterize heterogeneous migratory patterns. The DTM is a framework that keeps tabs on population movement and migration, aids in improving population dynamics analysis, and gives decision-makers and carers vital information in an emergency (Fig. 3.9). Since its creation in 2004 to monitor internal displacement in Iraq, DTM has been modified for use in 71 nations during times of violence, catastrophic events, complicated emergencies, and protracted crises. As of October 2020, there were 27.8 million IDPs, 20.1 million Returnees, and 5.1 million migrants.

Migrant integration has been more important on the international agenda since the 2030 Sustainable Development Agenda's aim to "leave no one behind," particularly migrants, was introduced. The topic of migrant integration, on the other hand, has a long history of polarization, and statistics on the subject is typically limited to high-income nations (Juran and Rachel 2016). At the individual, family, community,

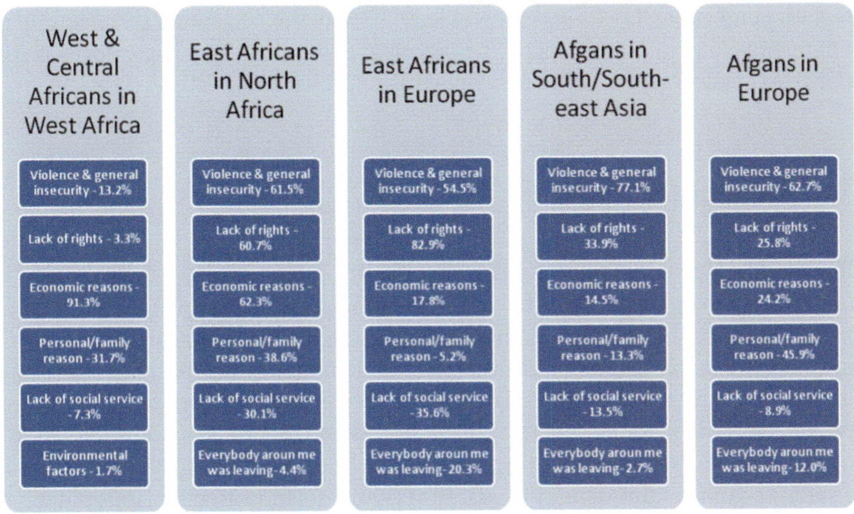

Fig. 3.9 Migration drivers and decision (*Source* 4Mi Graphic: Migration Drivers and Decisions, in Mixed Migration Review 2019, Mixed Migration Centre, Geneva (2019))

and national levels, integration happens in the public and private domains, across generations (Huddleston et al., 2015). Data on the incorporation of migrants includes a broad variety of subjects, including whether or not they are integrating into society's political, social, cultural, and economic spheres; prejudices they encounter; the ways in which policies impact their involvement; and public perceptions of immigration and migrants.

Most of the influx of people is due to migration, both internal and international (IOM 2015). Exoduses of migrant workers, both worldwide and domestic, have two main consequences for the countries and locations of origin: rising healthcare vulnerabilities (Zenner and Wickramage 2020) and socio-economic constraints. Beijing, Berlin, Brussels, Buenos Aires, Chicago, Hong Kong SAR, China, London, Los Angeles, Madrid, Moscow, New York, Paris, Seoul, Shanghai, Singapore, Sydney, Tokyo, Toronto, Vienna, and Washington DC were anticipated to house one in every five foreign migrants (IOM 2015). International migrants made up around 20% of the overall population in 18 of these cities. In some cities, the proportion of foreign-born people in the total population far outnumbers the global average (approximately 3.4% in 2015) (IOM 2015). In countries where the pandemic is spreading and where they are already afflicted, migrants—particularly those in low-wage jobs—may be more susceptible to the virus, but they also contribute significantly to the COVID-19 response by working in important regions. According to GMDAC evaluation determined by UN DESA (2020), World Bank (2020), WHO (2021), migrants from the 20 countries with the highest number of COVID-19 cases accounted for 31% of all international migrants as of 1 March 2021, and they sent approximately 37% of the total remittances to their place of origin in 2019. Of the 20 countries with the highest number of COVID-19 cases, at least 3.7% of the population was made up of immigrants, and in nine of them, the percentage of immigrants is greater than 7% (GMDAC assessment according to UN DESA (2020), WHO (2021).

3.6 Conclusion

Recent developments indicate that a significant number of people are moving to cities. More than 70% of people on Earth will reside in cities by 2025. The urban lifestyle will be the global economy's major engine, but it will also be the planet's most unsustainable source of unsustainability. Pollution, water scarcity, and social inequity are just a few of the issues that are looming on the horizon. The positive interpretation is that cities, as they have been throughout history, are the world's modernization research laboratory, and can thus be expected to generate solutions in a variety of spheres that define the quality of urban life: infrastructure, transportation, health care, education, leisure, safety, and so on. Traffic congestion, crime, economic stagnation, demographic segregation, and air pollution are all new difficulties for cities as a result of urbanization combined with persisting economic volatility and global climate change. Infrastructure has smartened up as a result of digitalization. The process of digitization has improved organizations' and firms' ability to work effectively, make the most use of their physical premises and energy, as well as oversee users, property, and processes proactively. In practice, smart governance is defined by a supply-oriented, technocratic approach to city management. The importance of technology in gathering data and providing information to smarten government operations and streamline urban system activities is highly emphasized in this process. With the introduction of IoT, cloud, blockchain, virtualization, robotics, and driverless vehicles—which have routinely supplanted people in tasks while boosting productivity and cutting costs—the Fourth Industrial Revolution got underway. Artificial Intelligence and deep learning will underpin the Fifth Industrial Revolution, completely replacing humans in production and managerial operations. The digitization of energy consumption, as shown by smart grids, will lead to an increase in infrastructure connectivity. Citizens and consumers will be able to make greater use of smart city space, services, and structures thanks to big data (3S). Horizontal platforms that offer a plethora of data, like city marketing, facilitate this approach. Information and communication technology will be used to send this content to each silo, enabling more efficient service administration.

References

Bull R (2016) Smart citizens for smart cities: participating in the future. Proc Inst Civ Eng Energy169(3):93–101

Cardullo P, Kitchin R (2017) Being a 'citizen' in the smart city: up and down the scaffold of smart citizen participation. The programmable city working paper 30, Maynooth, p 24

Fargues P (2020) Focus on operational data: the international organization for migration's displacement tracking matrix, and the mixed migration centre's mixed migration monitoring mechanism initiative. In: Migration in West and North Africa and across the Mediterranean, Edited Volume, Chap. 1. International Organization for Migration, Geneva, pp 6–20

GMDAC (2019) Migration data and the sustainable development goals. International Organization for Migration – Global Migration Data Analysis Centre (GMDAC)

Holubčík M et al (2012) Optimization of the production process of wood pellets by adding additives. Int J Energy Optim Eng 1(2):20–40

Huddleston T et al (2015) Migrant integration policy index 2015. Migration Policy Group, Brussels

International Organization for Migration (IOM) (2015) Conference on migrants and cities. IOM, Geneva

IOM Displacement Tracking Matrix (DTM) (2017) Methodological framework used in displacement tracking matrix operations for quantifying displacement and mobility. Global DTM Team, Geneva

Juran S, Rachel S (2016) The potential of the 2010 population and housing census round for international migration analysis. In: Conference background paper, improving data on international migration towards agenda 2030 and the global compact on migration, IOM, Berlin

Mixed Migration Review (2019) Highlights and trends. Mixed Migration Centre

Mutiara D, Yuniarti S, Pratama B (2018) Smart governance for smart city. IOP Conf Ser: Earth Environ Sci. https://doi.org/10.1088/1755-1315/126/1/012073

Nosek R et al (2017) Analysis of paper sludge pellets for energy utilization. BioResources 12(4):7032–7040

Rochet C (2015) Urban lifecycle management: a research program for smart government of smart cities. Institut de management Public et Gouvernance Territoriale, Marseille, p 33

The Worldwatch Institute Island Press (2016) Can a city be sustainable? (State of the world) Washington DC

United Nations (2018) Sustainable cities, human mobility and international migration: report of the secretary-general. UN, New York

United Nations Department of Economic and Social Affairs (UN DESA) (2018) World urbanization prospects: the 2018 revision, UN DESA, population division

United Nations Department of Economic and Social Affairs (UN DESA) (2020) Available at: https://www.migrationdataportal.org/themes/migration-data-relevant-covid-19-pandemic

WHO (2021). Available at: https://covid19.who.int/

World Bank (2020) Migration and remittances data. Available at: https://www.worldbank.org/en/topic/migrationremittancesdiasporaissues/brief/migration-remittances-data

Zenner D, Wickramage K (2020) Migration and health: a global public health research priority. BMC Public Health 20(1):1232–1240. https://doi.org/10.1186/s12889-020-09235-4

Chapter 4
Environmental Management, GeorbIS, and Smart Governance

4.1 Introduction

The way our governments run is only one aspect of our life that is impacted by modern technologies. Sensible administration in the areas of economic infrastructure, schooling, safety, transportation, and management of resources will assist democracies throughout the biosphere. Effective governance is a fundamental tenet of our societies and a great place to start when it comes to guaranteeing public welfare and state advancement. Nevertheless, people are decreasing trust in the system as a result of the limitations of good governance, such as deception, a lack of community participation, and discriminatory legislation. Because of this, effective governance may make the situation stronger by encouraging collaboration between businesses and citizens. Additionally, smart governance may now engage and coordinate with businesses and individuals in a technology-rich environment thanks to technological advances in information and communication technology (ICT) (Fig. 4.1). Additionally, smart governance can facilitate the more effective application of good governance principles.

Using modern technology and information and communication technologies (ICT) to create a translucent cooperative, long-term atmosphere that benefits citizens and government is known as "smart governance." There are four different models that can be used to create smart governments:

- Government-to-Citizen (G2C) Model: The administration engages with residents through communication outlets such as the newspaper, radio, television, and the Internet under this strategy. Furthermore, the government provides a forum for residents to prompt their assessments on government policies and programmes.
- Government-to-Business (G2B) Model: To facilitate economic growth, the government has discussions with businesses. Businesses may get firsthand knowledge of new laws, rules, taxes, and credit opportunities. Furthermore, governments encourage firms to conduct business online to save time and money, and businesses can obtain real-time data from the government.

G. S. Bhunia and P. K. Shit, *Urban Morphology and Sustainable Smart Cities*,
https://doi.org/10.1007/978-3-031-97473-1_4

Fig. 4.1 Conceptual framework of smart city

- Government-to-Government (G2G) Model: This establishes a framework for interaction between government and government institutions, departments, and agencies. The goal is to use ICT tools to unify all channels of government for a paperless, cronyism, and long-term system.
- Government-to-Employee (G2E) model: The government can communicate with employees and businesses using this paradigm. Additionally, a government database has the personal data of employees, particularly their social security numbers and information about bank accounts. Employee-related operations like payroll, bank loans, medical plans, and pension plans can also be completed online (Fig. 4.2).

4.2 Environmental Quality and Variability in Cities

Opportunities for economic and social benefits may be crucial drivers of urbanization and migration from rural areas to major cities (Chen et al. 2014). This would also put more strain on the environment and natural resources in metropolitan areas (Dwevedi et al. 2018). To secure the sustainable coexistence of our society, economy, and environment, the difficulties that accompany the rising population in cities require an immediate solution, which will necessitate a paradigm shift in city management (Gil-García et al. 2015). Smart city technologies have a vast, largely untapped potential for improving people's lives. By utilizing technology and data, smart cities aim to improve quality of life by implementing smarter decisions. Significant advancements in the environmental field are feasible, in addition to the advantages in terms of security, time, health, connection, employment, and cost of living. Smart city technology, such as energy optimisation, monitoring the quality of the air, and power, water, and garbage tracking, may reduce greenhouse gas emissions by 10–15%, solid waste

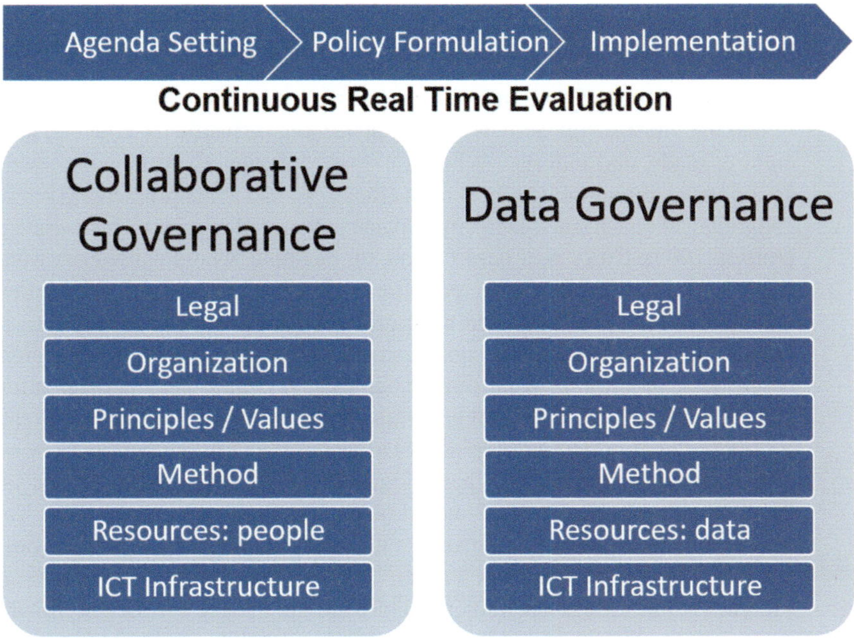

Fig. 4.2 Implementation smart city project through smart governance

generation by 30–130 kg per person annually, and daily water savings of 25–80 L per person.

To get these benefits, a city must have three tiers of intelligence, which build upon traditional physical and social infrastructure. In the technological basis, there are first connections of interconnected gadgets and sensors, including smartphones connected by high-speed networks of communication. The raw data is then transformed into alerts, insights, and actions using smart apps and data comprehension monitoring (Angelidou et al. 2018). Ultimately, successful data management and broad application uptake by enterprises, localities, and the general public encourage improved decision-making and behavioural changes. Innovative applications connected to trash, energy, water, and mobility make up the largest portion of environmental advantages. Real-time public transit information and building automation systems, for example, can reduce GHG emissions; improved air quality can be realized as a side effect of many energy-saving and mobility implementations; leakage identification and prevention can aid water conservation (Bakıcı et al. 2013); and digital traceability and payment for waste disposal can help reduce solid waste. The findings show that smart technology can enhance key indicators by 10–30% once implemented, and that adopting the current generation of smart city applications can efficiently assist cities reach 70 per cent of the sustainable development goals.

The goal of smart cities is to make them more sustainable. (1) Land and geography, (2) climate, (3) atmospheric pollution, (4) water resources, (5) energy, and (6) urban green space and biodiversity are the critical elements of sustainable cities in the World Cities Report 2016 and sustainable development goals (SDGs).

(i) ***Land and Geography***—Land will be one of the most important components in making cities sustainable, with ever-increasing demands for housing, agriculture, roads, and industry. When it comes to designing sustainable towns, knowing the topography of the land is crucial. To tackle the problems given by geographical constraints, each city's management will require its own strategy (Antrop 2005). Cities near desert areas, for example, would experience water scarcity and particulate matter pollution difficulties; as a result, such cities' management policies will need to prioritize effective water harvesting technology and particulate matter pollution control (Gill et al. 2008). Research-based data, monitoring data, remote sensing (RS) and Geographical Information System (GIS) data, and government documents will all be part of this massive collection. The development of a database, such as the World Overview of Conservation Approaches Technologies (WOCAT) global data, to analyse the economic benefits of SLM should be part of smart city management. The cost–benefit analysis of Sustainable Land Management Technologies has been aided by WOCAT (Giger et al. 2015).

(ii) ***Climate***: Future cities must implement measures that make cities more climate-resilient, according to SDG 11 of the sustainable development goals. To reduce air pollution, urban heating, and natural catastrophe susceptibility, these cities will require mitigation measures. Climate data from global and local organizations/institutes will be required to develop policies to mitigate the effects of climate change on cities. Climate data from the past and present is both important in planning, so climate data should include both real-time and long-term monitoring data on changing climatic conditions.

(iii) ***Atmospheric Pollution***: Large cities are constantly confronted with air pollution, which blankets them from time to time. Lack of regulation and foresight in transportation, industries, energy consumption, and waste generation are all major contributors to air pollution. Polluted air is known to have an impact on big populations all over the world. Each year, more than seven million people die prematurely as a result of air pollution, as per the World Health Organization (WHO) research (WHO 2016). The Central Pollution Control Board (CPCB) and the State Pollution Control Board (SPCB) should have enough monitoring stations, and the Air Quality Index (AQI) should be updated on a regular basis. Satellite data may also be used to learn more about atmospheric changes, western disturbances, and the areas adjacent to smart cities. Smog or acid rain, which can travel vast distances, is caused by many agents or pollutants combining with other entities in our environment. Air pollution can be dramatically lowered in smart cities by using green transportation, bicycle lanes, more green-belt areas, renewable energy, and discouraging the burning of municipal and agricultural wastes. The advantage of smart city people depends on the rapid collection of big data, modelling, analysis, and forecasts, as well as the conversion of that data into practical information.

(iv) ***Water Resources***—Water is one of the most essential natural resources, and its sustainability has become one of the most pressing concerns of the twenty-first century as a result of global climate change (Molden et al. 2007). According to studies, 25% of the world's river basins are drying up before reaching the oceans as a result of changing climatic patterns around the world. India is recognized to be losing its surface water resources at a significantly faster rate than other countries, and over-sharing of water has become an important component of international relations (Uprety and Salman 2011). Water provision is one of the most basic needs of peri-urban and urban populations; water scarcity can have disastrous consequences for a country's economy and living standards (Hunter et al. 2010).

The key obstacles that smart cities will encounter are the depletion of ground water tables, pilferage during supplies, and concerns of economic valuation and allocation for farmers and homes. It is critical for smart cities to strive for water sustainability and to explore various options for doing so. To conserve water resources, organizations such as the Central Ground Water Board (CGWB), the water resources department, the irrigation and flood control department, and the Ministry of water resources, river development, and Ganga rejuvenation should collaborate.

(v) ***Energy***—The entrance of a large metropolitan population will result in a significant increase in energy consumption for housing, lighting, transportation, and water management. India must meet the needs of growing urban population, and energy management may prove to be a critical factor (Ouedraogo 2013). The human development index is recognized to have a substantial association with per capita energy usage (HDI). India's goal is to generate 175 GW of renewable energy by 2022, primarily to meet the energy demand in smart cities (Aayog 2017), with the construction of 100 new smart cities and power to everyone. It is critical to promote electric vehicles for transportation, as well as the use of photovoltaic panels (PVCs) for street lighting and other routes such as commercial space. Another key function of smart cities could be to reduce energy usage. To conserve energy, the green construction concept should be encouraged (Susca et al. 2011). Increased use of asphalt for roads and other concrete buildings around the world has resulted in the establishment of urban heat islands, which are places where the average temperature is substantially higher than the surrounding area (UHIs).

(vi) ***Urban Green Space***—The importance of urban green space (UGS) in providing a variety of ecological services cannot be overstated. Natural woods, grasslands, marshes, parks, and gardens are all part of the UGS. Fisheries, agriculture, carbon sequestration, health benefits, tourism and recreation, pollination, air purification, wetland and water purification, urban cooling, and noise reduction are all examples of ecosystem services. The aforementioned ecological services are extremely beneficial in solving a variety of city-related issues.

UGS has the potential to drastically minimize the negative effects of air pollution on individuals. Ozone, sulphur dioxide, nitrogen dioxide, carbon monoxide, and particulate matter are all removed from the atmosphere by vegetation. Noise pollution will be a major issue in smart cities, but it can be effectively mitigated by the well-planned construction of urban green spaces. Noise pollution can be absorbed, reflected, and deviated by plants and trees. Wetlands will also aid in flood control, fisheries, and the buffering of municipal effluent (Chazdon 2008). The natural forest serves as an indigenous gene bank for plants and other species, making it useful in the restoration of degraded ecosystems as well as the creation of new urban green spaces. UGS planning and development must be based on scientific and social evaluation; haphazard and uneven UGS development may result in ecological or social issues. In planning, tools like Global Forest Watch, Wetland Atlas, remote sensing, and GIS will be useful. Published studies, forest departments, and local research organizations can all provide information on plants, animals, and micro fauna.

4.3 Disaster Risk Reduction and Sustainable Cities

Natural disasters such as floods, droughts, earthquakes, cyclones, forest fires, desertification, and insect infestation have a variety of effects on development. Infrastructure, lifelines, and vital assets are all damaged by disasters, culminating in human, economical, and environmental consequences. Rehabilitation necessitates finance, which frequently necessitates the diversion of funds intended for development and service supply. As a result, disasters stall development efforts by lowering available resources and disrupting planning. Disasters also reduce society's economic viability by aggravating poverty, impairing small company and industry operations, and interrupting critical lifelines for economic activity and customer service. Human capital is also lost as a result of disasters because of the casualties, accidents, and long-term trauma experienced by those who are afflicted.

Cities account for 60–80 per cent of energy consumption and 75 per cent of carbon emissions despite occupying only 3% of the Earth's territory. To date, the drive towards urbanization has been supported by growing environmental pressures and rising demands for essential services, infrastructure, jobs, land, and cheap housing, especially among the approximately 1 billion urban poor who live in informal settlements. Cities are highly susceptible to the effects of climate change and natural catastrophes due to their high influx of population, infrastructure, housing, and economic activities. Strengthening urban resilience is critical for avoiding personal, social, and economic losses, as well as protecting the environment and mitigating disaster risk and climate change (Fig. 4.3).

4.3.1 The Human-Natural Hazards Collision

Natural and man-made catastrophes are inextricably linked to development, which increases vulnerability to natural and man-made dangers. Development has an impact on the frequency and intensity of catastrophes, putting a rising share of the world's

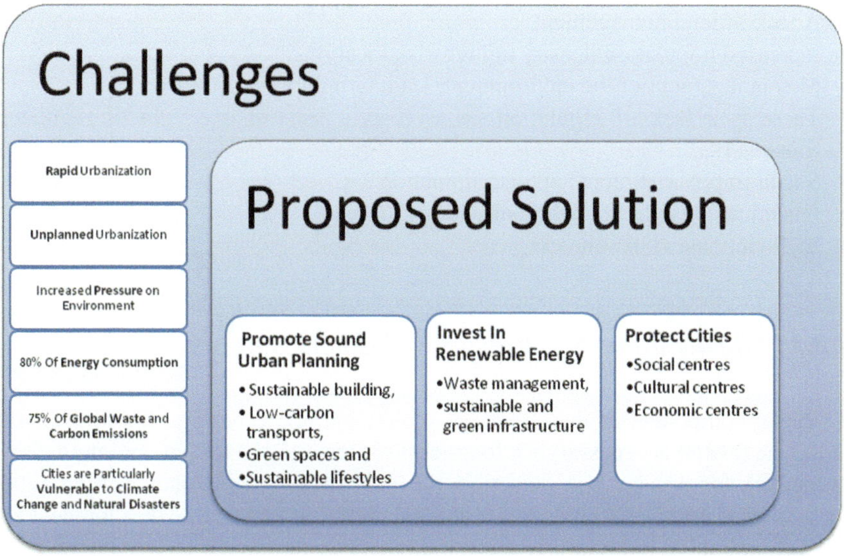

Fig. 4.3 Various challenges and solution for sustainable development of cities (*Source* Modified after UN Environment Programme)

population at risk. Rapid and unplanned urbanization has typified human growth in the developing nations during the previous half-century (ISDR 2002). The least developed countries account for ninety per cent of worldwide population increase. These countries lack the ability and resources to deal with the rapid urbanization. Between 1950 and 2000, the proportion of people living in urban areas in developing nations rose from less than 18% to over 40%. The rise is predicted to reach 60% by 2030. The top a hundred most populous cities have the greatest risk of calamity. At least one natural hazard affects more than three-quarters of these people. An average frequency of at least one major earthquake per fifty years is predicted for seventy of these communities.

4.3.2 Untenable Risk and Unsustainable Growth

Environmental deterioration (e.g., deforestation), overexploitation of natural resources (e.g., water), ecological disruptions (e.g., pollution), and social impoverishment (e.g., increased poverty) have all been linked to unplanned and ill-planned urbanization. Increased population density and inadequate building make the urban environment more vulnerable and socio-economic systems more vulnerable. Natural hazard vulnerability is frequently overlooked in land use and urban development approaches. In summary, disasters and innovation are linked by the following criteria:

- A lack of land management.
- Increased population density in high-risk locations.
- Mismanagement of the environment, leading in biodiversity loss.
- There is a lack of regulation, as well as a lack of oversight of regulatory frameworks.
- Social impoverishment and discrimination.
- Inadequately suited populations and institutions.
- Inefficient use of resources.

4.3.3 Cornerstones of Disaster Risk Reduction

A crucial component of the growth of society and the economy, and disaster risk reduction (DRR) is necessary for long-term success. As the first major international framework for disaster risk reduction, the Yokohama Strategy and Plan of Action for a Safer World (1994) recognized the link between sustainable development and DRR. Since then, this close relationship has been reinforced in key international pacts such as the Millennium Development Goals (MDGs), the Johannesburg Plan of Implementation (Johannesburg, September 2002), the "Hyogo Framework for Action (2005–2015)" and the "Future We Want" (Rio, June 2012), the Sendai Framework for Disaster Risk Reduction (Sendai, Mach 2016), and the 2030 Agenda for Sustainable Development (Sendai, Mach 2016). Announcing guidelines for reducing flood losses coincided with the 2004 World Water Day commemoration. This interagency publication, spearheaded by DESA, aims to provide decision-makers with a variety of solutions for decreasing flood-related damages. The sustainable development goals (SDGs) were announced by the United Nations in 2015, and they included catastrophe risk reduction and resilience in a variety of areas (Fig. 4.4). Sustainable development (SD) and DRR emphasize the benefits that communities and nations may experience by working together with a unified management plan that includes disaster risk reduction (DRR), climate change adaptation (CCA), and the promotion of sustainable future.

Human interference to lessen vulnerability reduces exposure to disasters, according to the worldwide disaster mitigation agenda. It is now commonly known that environmental deterioration, human development, and natural disasters are related. Most poor nations lack the people and financial means to adequately support disaster relief efforts. In a few circumstances where political will and pushed up exist, resources are insufficient to make a substantial difference. To emphasis on disaster prevention, governments, international and intergovernmental organizations, humanitarian organizations, the academic and professional community, and other stakeholders will need unprecedented human and financial resources. Disaster assistance should be seen as a challenging process that requires consistent, well-thought-out effort to be successful. The sustainability of programmes is critical for overcoming problems, surviving administrative turnovers, and beginning to effect socio-economic frameworks and disaster-related factors. With regulatory regimes focusing

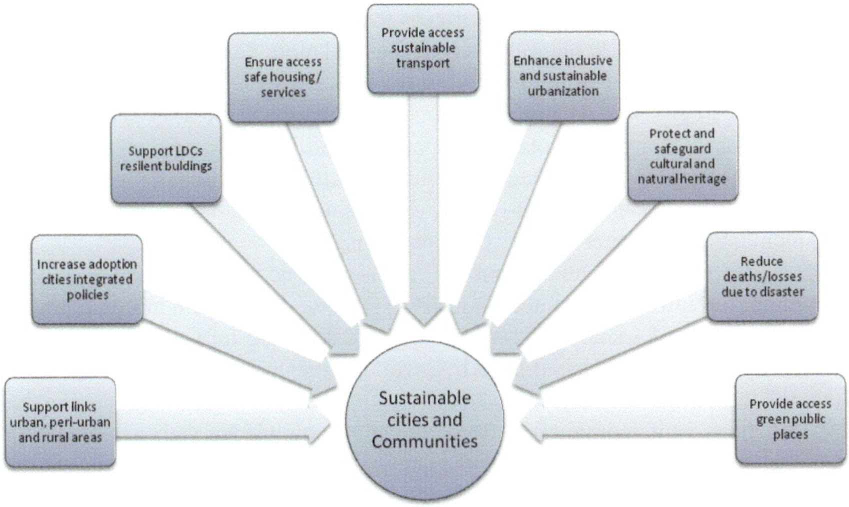

Fig. 4.4 Sustainable development goals and cities (*Source* Modified after UN Environment Programme)

on central decision-making and resources, local government disaster management activities are frequently given a lesser priority.

4.4 Digital Technology and Participatory Mapping in City Development

Informal settlements are an issue because they are unsanitary, dangerous, and lack proper social and technological infrastructure. Land is highly built-up due to a lack of planning and architectural expertise, with no public areas or proper streets. This dilemma will worsen as a result of ongoing urbanization. In her 2018 report to the General Assembly, UN Special Rapporteur Leilani Farha observed, "Currently, almost one-quarter of the world's urban population lives in informal settlements or encampments, largely in developing nations but increasingly also in the most affluent countries." The living conditions are appalling and unbearable. There are 520 million people living in informal settlements across Asia. Based on UN Habitat's proposal, states have committed to improving all informal settlements and providing enough housing for everyone by 2030 under Goal 11 of the Agenda for Sustainable Development. "Habitat 5.0" envisions a seamless merger of the digital and physical worlds. The latter relates to homes, resources, and active communities. The former refers to the application of digitalization concepts to spread and reuse existing architectural experience, which is compiled and arranged in what are known as "design patterns." Patterns are a well-known strategy that encourages learning from tried-and-true solutions. The blockchain supports deeper and more flexible pattern generation, combinations, sharing, community building, and licencing models. The benefit of this proposed strategy is that it can reach a larger number of beneficiaries and hence have a greater impact than prior, isolated campaigns by utilizing "disruptive innovations." The motto is "quality good enough" to stimulate self-construction of dwellings using appropriate ecological materials—safer, cheaper, faster, and more sustainable, thus complementing the larger aim of "Glocalization."

The Internet has altered communications and networks around the world, providing for information interconnectivity and scalability among and between different groups of society. The Mappiness and Ruimsig initiatives are similar in that they both address a social need by involving the community in rethinking and finding ideas, which are then depicted through maps. Both types of mapping in these two situations invite communities to define and acknowledge their experience in a geolocation. Participatory mapping, like many other social efforts, is built on trust and teamwork. Locals are the most knowledgeable about their communities and social issues, and they are prepared to donate their time and expertise to these sorts of initiatives. Local participants face the risk of having their collective intellect exploited, such as when ideas generated locally by local people are seized and sold for profit. Google Maps is a popular Application Program Interface (API) for mapping, with numerous usability options (for example, flexibility across multiple platforms), but it is not an open-source map like OpenStreetMap.org (OSM), which allows users to change the basic map itself.

Another significant conclusion of the Internet inquiry was that the procedure of spatial data infrastructure (SDI) development is now in the stage of digital data production in most of the searched countries and localities. Furthermore, the recorded data is frequently not yet provided through a geoportal or another decentralized online application. To accommodate this fact, the classification schema distinguishes between geospatial data availability and data accessibility. The outcome of an Internet inquiry into the evolution of geospatial data processing in the home nations of megacities reveals a wide range of results (Table 4.1). Some countries, such as Russia or Nigeria, are just starting to create an NSDI, but others are producing primary data (e.g., Iran, Pakistan) as well as secondary data (e.g., China, Japan). It may also be demonstrated that the development of an NSDI is well progressed geographical data via distributed web apps.

The concept of a smart city has been agreed upon by the industry, and it centres on innovative urban design with the goal of enhancing the six city components of human beings, economy, government, accessibility, way of life, and sustainability. ICT-based tools, such as publicly accessible databases and social media applications, can improve public involvement and assist in developing novel governance paradigms for smart cities. Furthermore, literature in the smart cities domain has emphasized the significance of data for smart governance (Kitchin 2014), the role of e-governance as a new vehicle for the emerging' smart city' (Paskaleva 2009), or IoT-based system for smart city development and urban planning using Big Data analytics (Rathore et al. 2016).

Table 4.1 Application of SDI in megacities

Country	SDI development status unknown	SDI master plan available	Primary spatial data available	Secondary spatial data available	Spatial data accessibility available
Argentina					√
Bangladesh		√			
Brazil					√
China				√	
Egypt	√				
France					√
Indonesia		√			
India					√
Iran			√		
Japan				√	
South Korea		√			
Mexico					√
Nigeria		√			
Pakistan			√		
Philippines			√		
Russia		√			
Thailand			√		
Turkey			√		
UK					√
USA					√

One of the most important aspects of smart city administration is citizen participation (Castelnovo et al. 2015). Citizen centricity is a conceptual model for governments to leverage ICT to improve citizen engagement in political debate and decision-making, hence effecting real change in public policy and governance (Chatfield et al. 2015; Reddick et al. 2015). When used by the public sector, social media networks play an important role in creating responsive governance. Bekkers et al. (2013) assert that social media tracking can facilitate enhanced responsiveness in policy making, collaborative creation with individuals, as well as procedures for governance where people engage and public organizations consider their ideas and suggestions. For example, by monitoring, analysing, and measuring social media activity, social media can help to identify the relevant stakeholders and establish greater collaboration, along with evaluate their perspectives (Falco and Kleinhans 2017). Open government data (OGD) is concerned with making public sector information available in open formats that the public may utilize and benefit (Kalampokis et al. 2011). It is critical to improve the interaction between government and the public in order to accomplish open governance, as well as to rely on information and communication technology (ICT) to arbitrate this process (Janssen et al. 2017).

4.5 Urban Heat Mapping

When the average temperature of a specific location exceeds the local norm, this is referred to as an urban heat island (UHI). Rapid urbanization in many developing nations has resulted in the extension of cities, the decline of vegetation cover, and the arid nature of cities with impermeable surfaces such as concrete and asphalt (UN Habitat 2017). The most significant component of any built environment are these transparent surfaces, and the planet's energy balance is impacted by their thermodynamic characteristics, which are different from those of natural land cover. They are also impenetrable, which means there is less moisture supplied to cool the area via evapotranspiration. This means that urban areas are hotter than the surrounding natural environment. The urban heat island consequence is the name given to this occurrence. We are already witnessing hotter and more frequent heatwaves as a result of climate change. These heatwaves have an influence on our community's health and well-being. High temperatures can have an influence on community health by increasing mortality, hospital admissions, and ambulance calls. High temperatures can have an impact on community health by swelling mortality, hospital admissions, and ambulance services.

Climate change is true, and experts agree that average worldwide air temperatures will rise significantly over the twenty-first century. Different representative concentration routes, or "emission scenarios," predict future global air temperatures based on anticipated greenhouse gas emissions and other factors encompassing a wide range of future events, including future population level, economic activity, and so on. Climate change is occurring, and scientists believe that average global air temperatures will rise significantly throughout the twenty-first century. Various representative concentration pathways, or "emission scenarios," forecast future global air temperatures based on projected greenhouse gas emissions and other aspects spanning a wide range of future events, such as future population level, economic activity, and so on.

However, geospatial data from Satellite Vu will be made to safeguard that these projects are truly ecologically sustainable, as it allows for detailed monitoring of a site as well as checking on energy and emissions at any time of day or night. Financial institutions can use this functionality to ensure compliance by allowing companies are becoming sustainable through their daily operations and have a real benefit on the universe around them. Satellite Vu data will be used to indemnify every building in a fund, rate the building based on its actual emissions, and make a comparison building by business class. Satellite Vu's role in new, smart cities will also grow in importance. The Internet of Things is intended to enter our cities like never before, and big data will allow governments and businesses to maintain capital more effectively.

The fundamental physics of UHI is the decrease in heat flux from latent sources and the rise in tangible heat flux. UHI is a serious environmental alteration caused by humans that puts human life in jeopardy. Increased morbidity and mortality, indoor/outdoor discomfort, air pollution, increased energy consumption and greenhouse gas emissions, impaired air and water quality, and expansion of energy poverty on vulnerable social groups during the hot months of the year are just a few of the UHI repercussions (Elmarakby et al. 2022). Thermal imaging is a technique for investigating UHIs that necessitates the use of a very accurate equipment, such as high-resolution infrared thermal cameras (FLIR T540). The thermal imaging technology camera receives radiation from the intended item in addition to atmospheric emissions (the radiation that surrounds the object) in order to provide the complete thermal states of the acquired object. An FLIR T540 Thermal Camera with 24° Lens, 464 348 (model: 79302-0201) was used to capture the thermal images in the four locations at the field measurement monitoring stations (FLIR, 2018). FLIR Tools® software was used to process the thermal pictures (Teledyne FLIR 2021). The temperature range between the average surface temperature (AST) and the surface temperature of the each urban material was calculated as part of the assessment (Elshater et al. 2019). The Statistical Package for the Social Sciences (SPSS) was then used to associate T with the proportion of material availability in order to investigate its impact on urban thermal warming (Kumar and Mullick 2012).

SHADEWAYS (https://www.shadeways.net/) allows users to share fine-grained heat data in order to create a low-temperature path and comprehend the data through a website and other materials. Current programmes, such as Google Maps, provide solely distance-based navigation advice. All navigation applications will be able to account for heat and shade with the proposed platform.

4.6 GeorbIS in Smart Resource Monitoring and Assessment

The word "GeorbIS" is a shortened form of "Geo," which refers to the earth, and "orbis," which indicates vision. "GeorbIS" is a precise platform for data gathering, coordination, and dissemination that was designed from the ground up to accommodate the 3D information. It enables precise transformation of large amounts of data into intelligent decision-making. The combination of Artificial Intelligence and geospatial data—which encompasses census, satellite, remote sensing, weather, mobile phone, hand-drawn, and social media data—allows for highly effective urban planning that can raise living standards both economically and environmentally friendly.

Satellite images provide data in various levels of detail, yet they are still underutilized. Buildings, roads, and other objects, on the other hand, can be difficult to recognize. Other simplified applications combine GIS and IoT data to provide realistic solutions to urban industrial concerns. Drones are being used by several firms to perform this work, but there are restrictions to this technique due to issues such as the size of the infrastructure, the absence of georeferencing, and technical glitches in constructing accurate 3D models/point clouds. In controlled airspace, the legal criteria for flying drones can be difficult to meet. Local laws may also confine their range by mandating a clear line of sight between the operator and the drone. Satellites take multi-angle images or video of the region of interest, which are then used to construct 3D digital models of the infrastructure and adjacent areas using advanced software. This allows us to compute the height and volume of buildings, as well as the height of neighbouring objects such as building, trees, pylons, and other structures. This enables us to evaluate the proximity hazards provided by these items to infrastructure. For example, assessing the height of a tree and its distance from a powerline or a building helps us to evaluate if the tree poses a threat or not.

The definition of a smart city, as stated in the National League of Cities study "Trends in Smart City Development," is "one that has established a technological framework that makes it possible it to acquire, interact, and evaluate real-time data to enhance the lives of its citizens." The modern smart city utilizes fully networked Internet of Things sensors and sensor nodes, smartphone applications, social media platforms, and smart devices to produce location-based data that optimizes a variety of aspects, including infrastructure, public safety, energy consumption, traffic jams, security, and emergency services. At the centre of the network is the application of geospatial technology, which lays out the parameters for gathering data, a language for assessment and decision-making, a mechanism for putting decisions into action, and a way to communicate with the public that is increasingly propelling smart cities.

As per the United Nations, 55% of the global populace resides in urban areas, and by 2050, that percentage is projected to rise to 68%. Eighty-seven per cent of Americans live in cities, and by 2050, that number is expected to increase to 87.4%. As per International Data Corporation, $80 billion was spent on smart cities in 2018, mostly spurred by initiatives in transportation, public safety, and energy, with spending expected to climb to $158 billion in 2022. The USA, which had trailed behind Europe and Asia in adopting smart city principles, led the world in spending on smart city technologies in 2018, with $22 billion, accompanied by China's $21 billion.

Another innovative and complementary function for geospatial intelligence (GEOINT) technology is emerging. During installation, commissioning, and maintenance, as well as a secure disposition, IoT cyberassets require meticulous chain of custody and geolocation tracking. Smart city networks, like traditional business networks, require strong access control policies and malware analysis. IoT sensors, unlike conventional networks, constitute a geographically dispersed population of network devices that cannot be securely stored in data centres. To avoid physical exfiltration and unlawful network access, a GEOINT-enabled security architecture detects the position of all network devices. GIS professionals with smart cities skills and expertise will be expected to deploy new and broader GEOINT technology as part of the security operation to fulfil the demands of tomorrow's cities.

4.7 Benefits

Citizen participation: Government and public participation now have a platform thanks to smart governance. ICT tools may be used by governments to communicate with citizens. Additionally, the public is able to offer comments and recommendations on the most recent initiatives, programmes, and policies. Their individual leaders, counsellors, and city officials would receive immediate input.

Accessing pertinent data: Smart governance has made it easier for citizens to obtain government data on finances, spending, and investments. All data should be publicly accessible, with the possible exclusion of vital information that may compromise individuals' security and safety. Because of the openness, there will be greater accountability for the government.

Better democracy: To avoid electoral fraud and abstention, eGovernment can employ identity and authentication methods. Additionally, voters will be able to fill out forms and validate themselves electronically with eID cards, making voting more comfortable and secure. Reliable online voting also affords voters confidentiality and privacy.

Financial and social inclusion: A significant portion of the global population is poor and illiterate. The utilization of financial services by this group is becoming increasingly challenging. As a result, smart governance may assist the impoverished by providing banking and financial solutions. Additionally, eIDs may be used to make a number of purchases for products and services.

Sustainable future: Smart governance will encourage people to make informed decisions. Improved policies for preservation of resources and safeguarding the environment, growth in neighbourhoods, citizen security, improved opportunities for learning and employment, and betterment of society will be made possible by the analytics generated by modern technology.

Involvement of the private sector: In the field of modern technology research and development, private enterprises have made tremendous strides. Organizations also have a good understanding of the strengths and limits of present technologies. As a result, commercial organizations can assist governments in preparing for successful technology adoption. Furthermore, private corporations may be interested in investing in government projects.

4.8 Conclusion

The future growth of smart cities will depend on how well big data is integrated and how we manage our resources for society's use. Our environment has to be maintained sustainably due to its significance and reliance on the services provided by ecosystems. Sufficient data is essential for carrying out the necessary evaluation, monitoring, feedback, and policy development that are all necessary for effective environmental management. A number of global, domestic, and local organizations and authorities are able to offer data on the smart city environment. The accomplishment of the Indian smart city objective is hampered by a variety of technological and environmental issues, such as the lack of regulations necessary for each smart city to be successful.

Clearly defining presently undervalued components like terrain, temperature, and green space and incorporating them into the smart city mission's standards for defining smart cities is the first hurdle. The second issue that need to be at the forefront of the advancement of smart cities is choosing how best to use resources and make future plans in light of the increasing dangers associated with climate change and natural disasters. To guarantee the success of India's Smart City Mission in future, the digital ecosystem—which comprises big data, ICT, and IoT—should be integrated with our natural milieu.

References

Angelidou M, Psaltoglou A, Komninos N, Kakderi C, Tsarchopoulos P, Panori A (2018) Enhancing sustainable urban development through smart city applications. J Sci Technol Policy Manag 9:146–169

Aayog N (2017) Draft national energy policy. Government of India: New Delhi, India

Antrop M (2005) Why landscapes of the past are important for the future. Landsc Urban Plan 70:21–34

Bakıcı T, Almirall E, Wareham J (2013) A smart city initiative: the case of Barcelona. J Knowl Econ 4:135–148

Bekkers V, Leonardi L, Spicer Z (2013) Contested spaces: the politics of public sector social media. Pub Manag Rev 15(9):1245–1258

Castelnovo W, Misuraca G, Savoldelli A (2015) Smart cities governance the need for a holistic approach to assessing urban participatory policy making. Soc Sci Comput Rev 34(6):724–739. https://doi.org/10.1177/0894439315611103

Chatfield AT, Reddick CG, Brajawidagda U (2015) Government surveillance disclosures, bilateral trust and Indonesia-Australia cross-border security cooperation: social network analysis of Twitter data. Gov Inf Q 32(2):118–128. https://doi.org/10.1016/j.giq.2015.01.002

Chazdon RL (2008) Beyond deforestation: restoring forests and ecosystem services on degraded lands. Science 320:1458–1460

Chen M, Zhang H, Liu W, Zhang W (2014) The global pattern of urbanization and economic growth: evidence from the last three decades. PLoS ONE 9:e103799

Dwevedi R, Krishna V, Kumar A (2018) Environment and Big Data: role in smart cities of India. Resources 7:64. https://doi.org/10.3390/resources7040064

Elshater A, Abusaada H, Afifi S (2019) What makes livable cities of today alike? Revisiting the criterion of singularity through two case studies. Cities 92:273–291

ElmarakbyE KM, Elshater A, Afifi S (2022) Tailored methods for mapping urban heat islands in Greater Cairo Region. Ain Shams Eng J 13(2):1–12

Falco E, Kleinhans R (2018) Digital participatory platforms for co-production in urban development: a systematic review. Int J Urban Reg Res 42(1):1–19

FLIR (2018) FLIR T540 professional thermal camera-FLIR systems. Available: https://www.flir.com/products/t540/

Gil-García JR, Pardo TA, Nam T (2015) What makes a city smart? Identifying core components and proposing an integrative and comprehensive conceptualization. Inform Pol 20(1):61–87.

Giger M, Liniger H, Sauter C, Schwilch G (2015) Economic benefits and costs of sustainable land management technologies: an analysis of WOCAT's global data. Land Degrad Dev 29:962–974

Gill SE, Handley JF, Ennos AR, Pauleit S, Theuray N, Lindley SJ (2008) Characterising the urban environment of UK cities and towns: a template for landscape planning. Landsc Urban Plan 87:210–222

Hunter PR, MacDonald AM, Carter RC (2010) Water supply and health. PLoS Med 7:E1000361

ISDR (2002) Living with risk: a global review of disaster reduction initiative, preliminary version. ISDR, Geneva

Janssen M, Matheus R, Longo J, Weerakkody V (2017) Transparency-by-design as a foundation for open government. Transform Govern People, Process Policy 11(1):2–8

Kalampokis E, Hausenblas M, Tarabanis K (2011) Combining social and government open data for participatory decision-making. Electronic Participation, 36–47

Kitchin R (2014) The real-time city? Big Data and smart urbanism. GeoJournal 79(1):1–14

Kumar S, Mullick SC (2012) Glass cover temperature and top heat loss coefficient of a single glazed flat plate collector with nearly vertical configuration. Ain Shams Eng J 3(3):299–304

Molden DK, Barker FR, De Fraiture C, Mati B, Svendsen M, Sadoff C, Finlayson CM (2007) Trends in water and agricultural development. In: Molden D (ed) Water for food, water for life: a comprehensive assessment of water management in agriculture. International Water Management Institute, London, UK

Ouedraogo NS (2013) Energy consumption and human development: evidence from a panel cointegration and error correction model. Energy 63:28–41

Paskaleva KA (2009) Enabling the smart city: the progress of city e-governance in Europe. Int J Innov Region Dev 1(4):405–422

Rathore MM, Ahmad A, Paul A, Rho S (2016) Urban planning and building smart cities based on the internet of things using big data analytics. Comput Netw 101:63–80

Reddick CG, Chatfield AT, Jaramillo PA (2015) Public opinion on national security agency surveillance programs: a multi-method approach. Gov Inf Q 32(2):129–141. https://doi.org/10.1016/j.giq.2015.01.003

Susca T, Gaffin SR, Dell'Osso GR (2011) Positive effects of vegetation: urban heat island and green roofs. Environ Pollut 159:2119–2126

Teledyne FLIR LLC (2021) Thermal analysis and reporting (desktop). FLIR Tools. Available: https://www.flir.com/products/flir-tools/.

Uprety K, Salman SMA (2011) Legal aspects of sharing and management of transboundary waters in South Asia: preventing conflicts and promoting cooperation. Hydrol Sci J 56:641–661

U.N. Habitat (2017) New urban agenda, UN Habitat III conference—United Nations sustainable development goals habitat III, Ecuador

WHO (2016) Ambient air pollution: a global assessment of exposure and burden of disease. World Health Organization, Geneva, Switzerland

Chapter 5
Smart City and Geospatial Technology

5.1 Introduction

While there isn't a single, agreed-upon definition of a "smart city," we can easily pinpoint the essential domains that are necessary to build more practicable, resilient, and adaptable urban environments: smart transportation, public administration, administration, and public goods and services like utilities, medical treatment, and educational opportunities. A geographical component, in combination with modern technology, plays a vital role in practically all of these fields. Utilize interactive web maps to arrange, arrange, and interpret pertinent data regarding public health (hospital infrastructure, illnesses propagation), road infrastructure (signs, potholes, and damaged roads), traffic (roadway construction, traffic delays, best travel routes, etc.), and excellent community behaviours (citizen initiatives, endeavours, and socioeconomic heritage).

The field of geospatial technology, usually referred to as geomatics, includes remote sensing, photogrammetry, visualization, Geographic Information Systems (GIS), geodesy, and the Global Navigation Satellite System (GNSS). The geospatial industry can be defined as "an information technology field of activity that obtains, manages, interprets, integrates, displays, analyses, or otherwise uses data focused on the spatial, temporal, and geographic environment," according to the US Department of Labour. With the use of latitude and longitude, a national coordinate grid, postal codes, or another method, geospatial technology gathers data from sensors placed around the city and systematically relates it spatially.

For many years, cities have used geospatial technology to enhance administration and amenities. Numerous government agendas can benefit from it, such as those pertaining to crime prevention, emergency preparedness, disaster relief, social services, health care, transportation, planning for urban areas, preservation of the environment, and building design and management. A variety of ICT and GIS technologies must be used at different phases to create smart cities. Cities are progressively providing their data openly available via open geospatial services (maps) that

G. S. Bhunia and P. K. Shit, *Urban Morphology and Sustainable Smart Cities*,
https://doi.org/10.1007/978-3-031-97473-1_5

reflect their policy. Virtual representations of all transactions and changes culminate in informed and engaged people. Indeed, the smart city concept aligns with advances in geospatial technology, which are leading to greater real-time data inputs, 3D visualization, and the capacity to follow development over time. GIS technology and geospatial analysis may clearly be used by governments and citizens to improve the quality of services. Citizens in Boston can use the Citizens Connect service to report municipal issues such as vandalism or damage to public infrastructure. Users can report problems through the website, a contact centre, or a mobile app. Every report is geotagged, forwarded to the appropriate agency, and resolved as quickly as possible.

Geospatial technology offers vast opportunities for infrastructure with the development of technologies like 3D modelling, LiDAR and additional terrestrial scanning, mobile mapping, surveying, location-based services, satellite data, high-resolution satellite images, and photogrammetry. The building information modelling (BIM) picture is incomplete without geospatial standards. BIM is a great deal more than a compilation of facilities management (FM) and computer-aided design (CAD) blueprints in two or three dimensions. The facilities and its extensive database must be linked to the property on which it is built and given access to owners and shareholders as a useful tool. Property borders, zoning, soil data, elevation, jurisdictions, aerial photographs, land cover and land use, and other geospatial information are all linked to and used in a BIM. Information about this will be of interest to customers, owners, financial institutions, realtors, rescue workers, repairers, tenants, safety inspectors, attorneys, emergency plans, and anyone working on nearby facilities.

5.2 Geospatial Technology—A Platform for Smart City

As urban areas become overcrowded and future growth momentum becomes extremely limited, the creation of new self-sustaining cities is emerging as an interim approach to these issues. These new autonomous cities are powered by a network of sensors, cameras, portable electronics, and data centres that enable mechanization and real-time comprehensive city surveillance and administration. These new self-sustaining cities, also known as smart cities, are built urban regions that excel in numerous important areas such as economy, environment, energy efficiency, mobility, governance, people, and living environment to produce economic growth and development and high quality of life. A key factor in determining success is the need for a single technological platform to enable the integration, cooperation, and synergistic functioning of many players in the ecosystem of smart cities. A centralized information system based on Geographical Information System (GIS) delivers an IT framework for integrating not only all stakeholders (Fig. 5.1), but also all aspects of smart city activities, from conception to establishment and improvement.

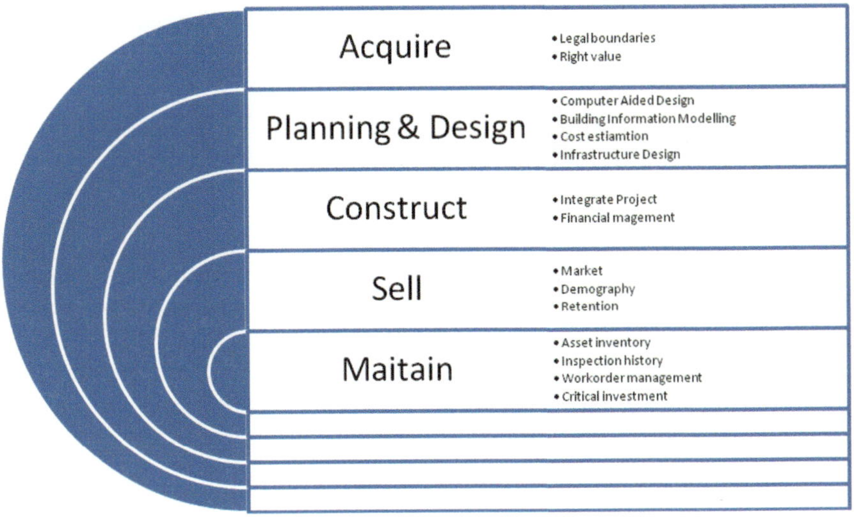

Fig. 5.1 GIS-based centralized information system

5.3 Framework of Geospatial Technology for Smart City

5.3.1 Site Selection–Land Acquisition

GIS can collect and synthesize many sorts of data to aid in decision-making, as well as providing high-quality visualization apparatuses that can aid in empathetic and decision-making with regard to site assessment, valuation, and eventual decision. Geographical information such as land usage, soil fertility, ground water depth, soil carrying capacity, and susceptibility to natural disasters like earthquakes and floods can be examined by real estate organizations in order to determine the accurate property assessment. The benefits of one site or location over another can be assessed by analysis, mapping, and modelling. This can also be used to determine suitable market-linked remuneration for owners based on investment strategies, as well as to plan restoration and evacuation.

5.3.2 Environmental/Legal Compliance

GIS facilitates the completion of regulatory responsibilities more quickly and easily by providing a standardized platform for interacting with regulators and the public. The instantaneous linking of pre-existing data to a compliance workflow guarantees conformance. Additionally, the quick creation of reports that clearly demonstrate how standards and building bylaws are followed can be facilitated by GIS-based graphical outputs.

5.3.3 Planning, Design, and Visualization

At every stage of the process, from project idea to site research, design requirements, stakeholder collaboration and engagement, design production, simulation, and assessment, geodesign will be a crucial foundation for conceptualizing and developing smart cities. Planners can use GIS to combine data from a range of sources, such as roads, sewers, and drinking water, and do spatial queries and planning. Thousands of miles of overhead and underground networks can be managed and mapped by utilities.

5.3.4 Construction and Project Management

GIS, when combined with project management and finance software, includes an overview of projects and their present situation, as well as aids in performance tracking. Using GIS, a user can manage all crucial project information—including soil data, geotechnical studies, scheduling, ecological parameters, engineering drawings, project diagrams, inventory, and asset control.

5.3.5 Sales and Marketing

City planners can use GIS to entice potential firms by generating useful sales tools and marketing reports that showcase a new location's or forthcoming development's commercial viability. Residents benefit from GIS because it provides a visual depiction of all the data that influences a property's popularity and worth, providing them a considerably more realistic picture of a property's compatibility to their demands.

5.3.6 Facility Management (FM)

By creating comprehensive information that aids in smarter resource allocations, the scalable basis for improved facilities management is provided by a GIS-based system of data. The present property management system can be integrated and extended with GIS. By transmitting and compiling the geometries and statistical information of the multiple BIM and/or CAD files necessary to accurately represent the urban spaces into a GIS, the effectiveness and benefits of BIM can be extended, strengthened, and connected geographically to other pertinent site, neighbourhood, municipal, and regional data.

5.3.7 Operations and Reporting

In addition to tracking and evaluating assets across space and time, GIS can provide insights by presenting data as simple reports and maps. It enables you to design an activity's perspective using the most recent geographic data that is displayed in a web map or web service, which can contain lists, charts, gauges, maps, and others. To address the stakeholder needs on various parts of the operation, many operation perspectives can be established. GIS gives townships more control over operations and has a favourable influence on the bottom line by allowing them to integrate different data sources into a single operational image of all facilities.

5.4 Uses of GIS in Smart City Project

The implementation of smart city projects involves several stages, including the creation of an urban digital model, data collection via digital devices, data processing, dynamic data presentation, and system control.

5.4.1 Urban Digital Database

The creation of an urban computer model that defines the elements of the urban built and natural environments is the first stage in the adoption of innovative city initiatives. The digital model gives geolocation and features for each urban unit (attributes). Building information modelling (BIM) is used to describe "vertical components" such as buildings, whereas GIS is worn to create a digital model of urban "horizontal components" including urban networks, transportation amenities, and natural environment (Fig. 5.2). A powerful tool for building an urban digital model using georeferenced data and presenting that data in an approachable way is produced by the merging of GIS and BIM.

5.4.2 Integration with IoT/Digital Devices

This category includes computer programmes that use sensors for monitoring urban networks and infrastructures. When GIS and BIM are integrated, a potent tool is created that may be utilized to create an understandable urban digital model utilizing

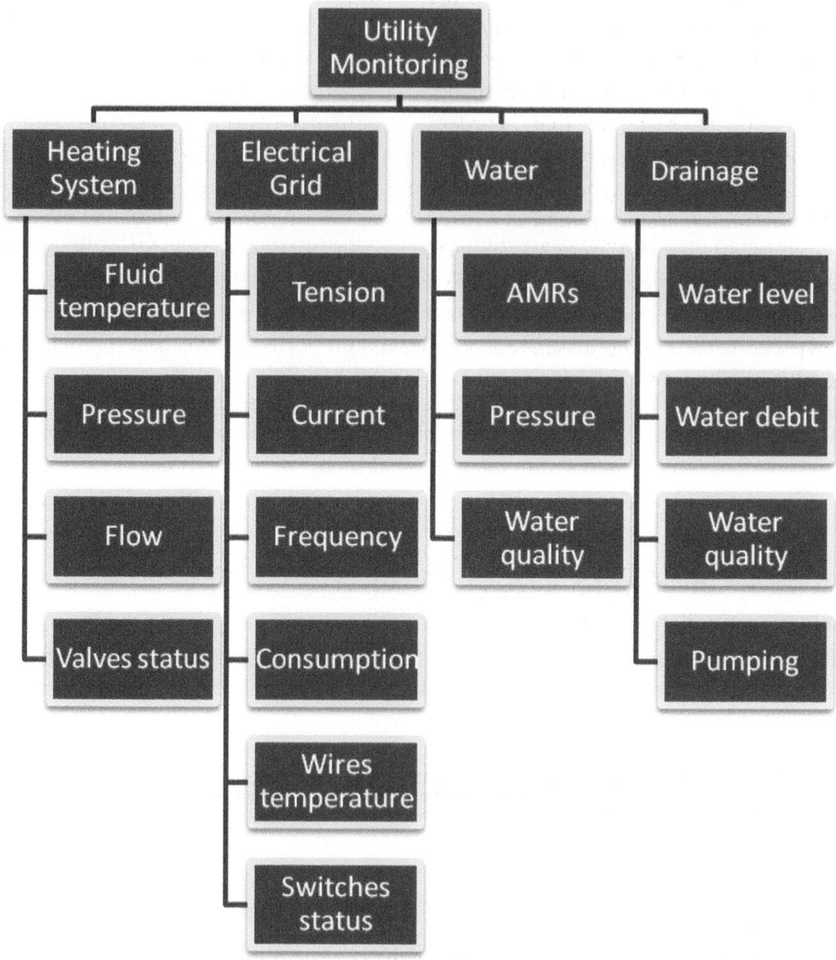

Fig. 5.2 Utility mapping and monitoring through GIS in city

georeferenced data. This category includes sensor-based information systems for keeping an eye on urban networks and infrastructure. Water quality monitoring sensors assess the turbidity, pH, chlorine, and conductivity of the water; automated metres readers (AMRs) track the amount of water used; pressure measuring tools track the water pressure. The drainage system uses sensors to monitor the water's flow and level as well as its general characteristics (pH, temperature, turbidity, etc.) and the functioning of the pumping apparatus. It makes it possible to automatically identify floods and pumping mechanism issues. There are sensors in the electrical grid that can identify frequencies, current, and electrical excitation. It makes it possible to identify electrical grid problems in a timely manner. The monitoring system, including its features and sensor states, may be seen using GIS. It also allows information to be visualized on GIS maps, both historical and present.

Global Positioning System (GPS) provides the foundation for developing smart solutions for smart cities. In order to provide real-time location-based services, GPS is essential. In resource allocation and municipal administration, real-time location sharing is critical. It aids with traffic management, crime prevention, product and service delivery, and disaster planning, among other things. GPS provides the foundation for developing smart solutions for smart cities. In order to provide real-time location-based services, GPS is essential. In resource allocation and municipal administration, real-time location tracking is critical. It aids with traffic management, crime prevention, product and service delivery, and disaster planning, among other things.

5.4.3 Integration Between GIS and Data Science

In order to improve the security, efficiency, and dependability of municipal services, the smart city project concentrates on building an integrated ecosystem that transforms both historical and current information into operational information. Engineering, management, and safety software for urban systems, as well as advanced digital technologies like artificial intelligence (AI), are all part of the analytic environment. GIS offers functionality for (i) geospatial data analysis (distance and directional investigation, geometrical processing, grid models), (ii) spatiotemporal analysis (spatial query, advance query, measuring tool), (iii) spatial statistics (spatial autocorrelation and egression), (iv) surface analysis (surface form and process flow diagrams, gridding and interpolation methods), and (v) location analysis (reactive routing calculation, capacity planning, surveillance system) in smart city.

GIS is critical to the smart city ecosystem because it integrates all parts of city planning and management and then creates a single operating image. With so many stakeholders participating in the integration, collaboration, and synergistic functioning and operation of a smart city, GIS data is essential for converting smart city ideas into action.

5.4.4 Interactive Data Visualization

Users can interact with the smart city's elements and participants in a user-friendly environment thanks to dynamic data visualization. This immersive environment is created using webGIS services. HTML popups allow visitors to view web-based media, including graphics linked to via URLs. Urban features and sensor maps can be visualized in the interactive GIS graphic environment. These maps can be used by users and administrators to obtain both static and dynamic data on urban systems, as well as to check for updates (Fig. 5.3).

Fig. 5.3 Geospatial in making cities smarter

5.5 Usefulness of Aerial Data in Smart City

An emerging technology that is driving the geospatial industry ahead is unmanned aerial vehicles (UAVs), sometimes referred to as drones. Microprocessors, sensors, wireless communication devices, and cameras are all built into UAVs. The road network is the most critical element to precisely map in any city. They are mostly employed in the city for transportation and traffic regulation. Second, most of the city's utilities, such as water and sewage lines, OFC/telephone cables, gas pipelines, and electric cables, are laid alongside the highways. Most importantly, the city's road network connects all of the city's properties. The road system's length in major cities is expressed in thousands of kilometres. As a result, in any smart city project, speedy and precise mapping of these road networks, as well as roadside utilities, is critical. There are some regions in the city where both of the above-mentioned data collection methods may not be practical, such as in unreachable or prohibited areas. This includes defense installations within the city, sensitive areas such as airports and nuclear research facilities, as well as sparsely populated hilly or steep regions and other similar zones. For such locations, high-resolution satellite images (HRSIs) with a resolution of around 0.5 m on the ground can be employed. Alternatively, if required, stereo pairs can be employed to create the topography and DEM of such locations.

The US Department of Defence first employed them for a specific reason (Vodák et al. 2021). They are presently being integrated into smart cities' variable area concepts, helping to achieve specific objectives (Mohamed et al. 2020). Unmanned aerial vehicle technology offers several benefits, including increased productivity, reduced expenses, adaptability, and a superior degree of consistency. The data transfer rate, which is critical for smart city ideas, is a competitive advantage of mobile drones. Drones can help with the building and management of smart cities in a variety of ways. Traffic monitoring, planning and progress monitoring of critical infrastructures, security, transportation, and mapping and surveying for frequent monitoring of development activities are only a few of the key uses. Experts are employing UAVs to collect, aggregate, analyse, and distribute highly precise and comprehensive data in most smart cities across the world.

- Drones equipped with advanced high-resolution cameras and post-processing techniques were used to conduct an aerial survey, which significantly decreased the time, money, and effort involved for surveying. Drones can also be used to build transmission lines, power grids, and power plants, as well as to evaluate them.
- Real-time traffic control and monitoring for congestion prediction and resolution, efficient parking spot detection, and development of traffic simulation models. Since they can be deployed quickly, drones can be useful for responding to emergencies, security, regulating crowds, directing traffic, parking, and weather evaluations. Over stationary sensors, drones can provide far richer and more usable data to city planners and municipal administrations. Citizens will benefit from the more versatile, cost-effective, and efficient gathering of information, as well as the distribution of municipal and other government services.
- Transport of defibrillators and emergency aid using an unmanned aerial vehicle serving as an ambulance.
- Police in the UK have used a drone to catch a car thief, and unmanned aerial vehicles can also track crowds using anomalous motion detection or live video streaming. Drones are being used by Canadian law enforcement authorities (for example, the RCMP) to carry out their missions. Additionally, certain Ontario roads are being monitored by drones to track high-occupancy vehicular traffic. However, because of the enormous volume and kind of data that drones may acquire via cameras and sensors, privacy and data protection concerns are one of the major roadblocks to their incorporation into cities.
- Drones can be seamlessly debuted to glean data in areas where conventional methods are difficult and complicated, such as monitoring the development of new buildings or investigating ageing infrastructure. Real-time data on construction locations and development schemes helps both the corporate and public sectors achieve their cost-cutting goals.
- A city's efficient safety and security system is a priority for every city or community, not just smart cities. During any event, most cities' police or private entities employ drones to keep an eye on the crowd. Smart patrolling is made easier with the use of unmanned aerial vehicles (UAVs). UAVs will also help smart cities become safer places to live by integrating mobile applications, wireless networks, and forensic mapping software.
- Bridges, pipelines, electrical lines, and buildings will be inspected as part of the Key Infrastructure Survey.

- The management of restricted resources, such as agricultural support, water distribution, emission assessment, and pollution control, is essential for achieving sustainable development.
- Travel utilizing a drone to film with a camera or to act as an internet-based guide offers innovative ways to explore and document new places.
- Geodetic work for remote sensing, 3D modelling, and GIS support; drones will aid in the transition to a 5G network by participating in network construction and locating dead zones. Consequently, a 5G network will allow thousands of drones to operate simultaneously, offer faster connection speeds, and allow drones to fly farther without being impacted by network fluctuations.
- Natural disasters, such as fires, earthquakes, and floods, are predicted and solved in urban ecology. UAVs can adequately assess the situation since, in some cases, they can access locations that humans cannot.
- Deliveries, wireless communication, and unmanned aerial vehicle taxi services, as well as data collection using drones and WSNs, are all possibilities (Wireless Sensor Network). In smart cities, drones can meet the overwhelming need for easy and accurate distribution. As a result of the COVID-19 epidemic, e-commerce has reached new highs, and retailers and consumers are deeply concerned about product delivery costs and timeliness. Drones can transport medical supplies as well as consumer goods. Drones can convey precise GPS location information for package tracking, along with reduce traffic congestion and gas emissions in urban areas, in creating and delivering items quickly.
- Drones can properly collect air temperature, atmospheric pressure, humidity, carbon monoxide (CO), carbon dioxide (CO_2), sulphur dioxide (SO_2), ozone (O_3), nitrogen dioxide (NO_2), and ammonia when used to measure air pollution levels (NH_3).

Drones are used in photogrammetry to transmit information on earthquakes, volcano or flood activity, green area recognition, and so on. UAVs are appropriate for usage if they are doing a task that is hazardous to humans or can be done more economically (Yadav et al. 2019).

The 3D Mobile Light Detection and Ranging (LiDAR) Technology might be used to map all of the city's available roads with a $\pm$ 5 cm precision. On a daily basis, we can record roughly 150 to 200 kms of road networks. Mobile LiDAR data can be acquired at speeds ranging from 10 kms per hour to 80 kms per hour. On city roads, the ideal vehicle speed is usually 30 to 40 kms per hour. With the use of dense 3D point-cloud information, 360-degree perspective camera data, and Global Navigation Satellite System (GNSS)/IMU data, Mobile LiDAR runs may effortlessly gather around 150 m of corridor, or 75 m on either side of the run. On city highways with medians, grade separators, and flyovers, numerous runs are frequently needed to accommodate high-quality 3D point-cloud data.

5.6 Future Strategy for Sustainable Urban Landscape

GIS technology has had a significant impact on public administration over the last few decades, enhancing fundamental record keeping and data management as well as automating a wide range of geospatially related operations, especially mapping (Ajmar et al. 2015). The majority of these enhancements have been made to administrative systems that are concentrated on certain areas of operation. We're going to enter a period where geospatial (and other) data about cities is exploding. This information is progressively being used to solve the mounting problems that cities are facing today. Managers and policymakers are looking for ways to effectively utilize digital data in order to improve decision-making and government administration (Robinson et al. 2017).

Consequently, GIS is increasingly recognized as an effective tool that local governments may use to accomplish these goals. Its way to mobilize, integrate, analyse, and visualize very massive and complicated data has made it a necessary foundation for the creation of future urban sustainability. GIS systems have generally relied on database-centric methodologies to produce information products like maps, reports, and views from centralized database management systems (Delgado et al. 2019). The notion of global enterprise solutions produced by interconnecting departmental platforms started to emerge over the last era, owing principally to the introduction of quicker processing and networks. For constantly merging data from numerous distributed databases, this idea relied on services-oriented paradigm.

Cities are progressively implementing designs of disseminated data mashups (both tabular and map). This aids in the unification and integration of data from a variety of sources within the organization and outside (van Steen and Tanenbaum 2016). The web and web services make this new work pattern possible. While it does not completely eliminate the requirement for proper traditional database design, it does provide a far more flexible foundation for designing and deploying GIS programmes. It is truly assisting in the realization of corporate GIS's data integration objective.

More data on cities is now becoming available instantly. Massive networks of stationary and mobile devices are being built to detect and quantify everything that moves or occurs (Jang-Jaccard and Nepal 2014). Traffic, utility practice, the environment, and smart building data are all provided as APIs. Crowd sourcing will eventually be implemented as an information source, allowing residents and city staff to contribute their findings and interpretations (Papadopoulou and Giaoutzi 2014). Data will eventually become able to be deliberately linked and harmonized thanks to functional data that is service-enabling (data kept in corporate processes). Cities in the future will be considerably smarter (Nevoa and Kotlarsky 2020). Everything will be monitored in real time and in great detail thanks to the use of advanced sensor arrays. Integrating mountains of real-time data and transforming it into comprehensible and useful will be made possible in large part by GIS. It will improve several applications, involving as land use and transportation planning, architectural design, and environmental quality assurance. Better decisions will be formed as a consequence, and collaboration will improve as well as productivity will rise.

A network of sensors, cameras, wireless devices, and data centres form the main infrastructure in smart cities, allowing civic authorities to perform important services more quickly and efficiently (Hancke et al. 2012). Smart cities are also far more environmentally friendly, as they employ sustainable materials in their construction and require less energy (Belli et al. 2020). The establishment of a powerful network of communication that links all companies, individuals, and interactions between the national and local government levels is made possible by the effective use of innovation. It also helps to create an effective transportation management system and better healthcare facilities.

This strategy will benefit all governments, regardless of their size or location. It provides solutions that span multiple disciplines within a government in order to boost the entire government's operations. Consider government to be a platform. The foundation of building smart communities is collaborating closely with governments and professional trade groups to fully comprehend government requirements. GIS is already used for this, and it will be used in the future to convert data from local governments into knowledge that cities can utilize to enhance services and improve the standard of life for their citizens. GIS specialists will serve to improve this development by creating an unified, translucent, and complete science-based architecture that will aid in the evolution of our cities into smarter and more sustainable future smart cities.

References

Ajmar A, Boccardo P, Disabato F et al (2015) Rapid mapping: geomatics role and research opportunities. Rend. Fis. Acc. Lincei 26:63–73. https://doi.org/10.1007/s12210-015-0410-9

Belli L, Cilfone A, Davoli L, Ferrari G, Adorni P, Di Nocera F, Dall'Olio A, Pellegrini C, Mordacci M, Bertolotti E (2020) IoT-enabled smart sustainable cities: challenges and approaches. Smart Cities 3(3):1039–1071. https://doi.org/10.3390/smartcities3030052

Delgado JA, Short Jr NM, Roberts DP, Vandenberg B (2019) Big Data analysis for sustainable agriculture on a geospatial cloud framework. Front Sustain Food Syst 16 July 2019 | https://doi.org/10.3389/fsufs.2019.00054

Hancke GP, Silva Bde C, Hancke GP Jr (2012) The role of advanced sensing in smart cities. Sensors (Basel) 13(1):393–425. https://doi.org/10.3390/s130100393. PMID: 23271603; PMCID: PMC3574682

Jang-Jaccard J, Nepal S (2014) A survey of emerging threats in cybersecurity. J Comput Syst Sci 80(5):973–993

Mohamed N, Al-Jaroodi J, Jawhar I, Idries A, Mohammed F (2020) Unmanned aerial vehicles applications in future smart cities. Technol Forecast Soc Chang 153:119293

Nevoa D, Kotlarsky J (2020) Crowdsourcing as a strategic IS sourcing phenomenon: critical review and insights for future research. J Strateg Inf Syst 29(4):101593

PapadopoulouCA,Giaoutzi M. 2014. Crowdsourcing as a Tool for Knowledge Acquisition inSpatial Planning. Future Internet 2014, 6, 109–125; https://doi.org/10.3390/fi6010109

Robinson AC, Demšar U, Moore AB, Buckley A, Jiang B, Field K, Kraak MJ, Sluter CR, Camboim SP (2017) Geospatial big data and cartography: research challenges and opportunities for making maps that matter. Int J Cartography 3(sup1):32–60. https://doi.org/10.1080/23729333.2016.1278151

Yadav D, Choksi M, Zaveri MA (2019) Supervised learning based greenery region detection using unnamed aerial vehicle for smart city application. In: Proceedings of the 2019 10th international conference on computing, communication and networking technologies (ICCCNT), Kanpur, India, 6–8 July 2019

van Steen M, Tanenbaum AS (2016) A brief introduction to distributed systems. Computing 98:967–1009. https://doi.org/10.1007/s00607-016-0508-7

Vodák J, Šulyová D, Kubina M (2021) Advanced technologies and their use in smart city management. Sustainability 13:5746. https://doi.org/10.3390/su13105746

Chapter 6
Role of Urban Morphology in Smart City Management

6.1 Introduction

Urban morphology—the study of the form, structure, and layout of urban spaces—plays a pivotal role in shaping the physical and functional aspects of cities. In the context of smart city management, understanding urban morphology is crucial for designing cities that are not only technologically advanced but also sustainable, livable, and resilient. Scholars such as Conzen (1960) and Lynch (1960) provided foundational frameworks, focusing on town plans, building forms, and land use patterns. These early works laid the groundwork for understanding how urban structures influence human activities and urban functionality. The spatial arrangement of buildings affects energy distribution and consumption. Studies have shown that well-planned urban morphologies contribute to reduced energy footprints (Steemers 2003).

The integration of urban morphology into smart city planning involves analysing the spatial arrangement of buildings, streets, public spaces, and infrastructure to ensure seamless functionality. Urban forms influence critical aspects of a city's operation, such as transportation efficiency, energy consumption, environmental quality, and social interactions. For instance, compact urban forms can promote walkability, reduce traffic congestion, and optimize resource utilization, aligning with the goals of smart city initiatives.

Smart city management leverages cutting-edge technologies like Geographic Information Systems (GIS), Internet of Things (IoT), and Artificial Intelligence (AI) to monitor, analyse, and improve urban systems. Studies such as those by Angel et al. (2016) emphasize GIS's role in visualizing and analysing spatial patterns, helping planners optimize urban layouts for better mobility and resource allocation. Urban morphology serves as the foundation for deploying these technologies effectively by providing insights into spatial patterns and urban dynamics. For example, GIS-based analysis of urban forms enables planners to optimize public transportation routes, design green infrastructure, and mitigate urban heat island effects. Batty (2013)

G. S. Bhunia and P. K. Shit, *Urban Morphology and Sustainable Smart Cities*,
https://doi.org/10.1007/978-3-031-97473-1_6

highlights how urban systems are interconnected networks, where spatial patterns directly affect urban dynamics like transportation, energy distribution, and social interactions.

Emerging trends in the field include the application of advanced computational tools such as digital twins, which provide real-time 3D representations of urban morphologies for dynamic analysis and planning (Ramirez Lopez and Grijalba Castro 2021). Additionally, the growing emphasis on sustainability has spurred research on integrating morphological studies with the United Nations' Sustainable Development Goals (SDGs). Moreover, urban morphology contributes to achieving sustainability and resilience in smart cities. Thoughtful planning of urban layouts can enhance disaster preparedness, improve drainage systems, and ensure equitable access to amenities. By focusing on the interrelationship between built environments and human behaviour, urban morphology helps create cities that prioritize citizens' well-being and environmental stewardship.

6.2 Urban Morphology: Fundamentals and Framework

Urban morphology is the study of the physical form and structure of cities, focusing on how urban spaces evolve and how various elements interconnect to shape the built environment. It delves into the relationships between urban forms, spatial organization, and societal processes. The key elements of urban morphology include physical elements such as urban blocks, streets, and open spaces, which govern the layout and movement within the city. Structural elements like street patterns, building heights, and urban edges define the city's hierarchy and organization. Social and cultural elements reflect the community's needs and historical influences, shaping land use, identity, and public spaces. Additionally, natural and environmental elements such as topography, hydrology, and green spaces influence urban development and sustainability. Understanding these fundamental components provides a framework for analysing cities, guiding urban planning, conservation, and resilience efforts.

6.2.1 Physical Elements

Physical elements of urban morphology refer to the tangible, built structures that define the physical layout and spatial organization of a city. These elements include urban blocks, streets, roads, intersections, and open spaces, all of which determine how people move, interact, and experience the urban environment. The street network, for example, influences connectivity and accessibility, while the design of streets and public spaces contributes to the city's aesthetic and functional qualities. Urban blocks, or building lots, shape the arrangement of buildings and determine land use patterns, such as residential, commercial, or industrial zones (Fig. 6.1). Additionally, public spaces like plazas, parks, and squares serve as social hubs, enhancing

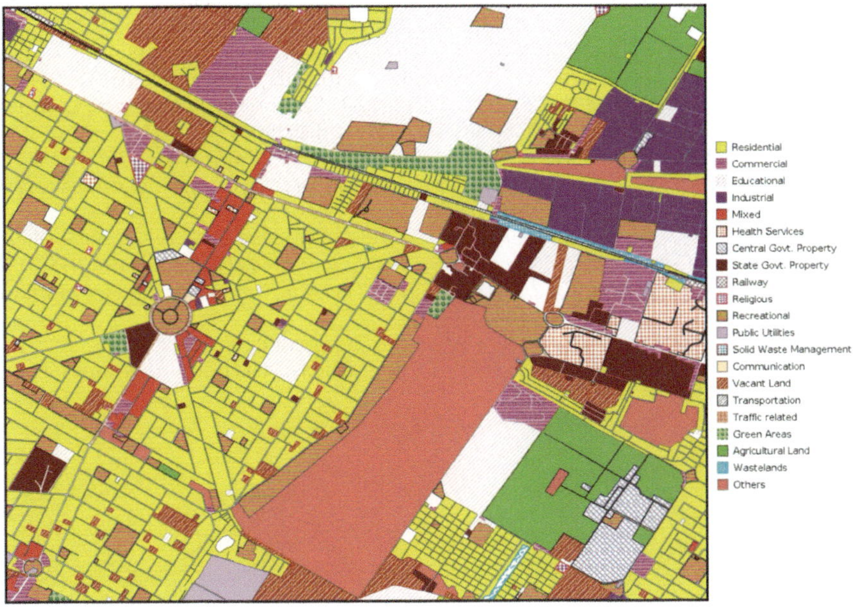

Fig. 6.1 Sample view of physical elements in urban centre

the livability and vibrancy of the city. These physical elements, together, create a framework for understanding how cities are organized, how spaces are utilized, and how urban development responds to both human needs and natural features.

6.2.2 Structural Elements

Structural elements of urban morphology refer to the underlying frameworks that shape the organization and configuration of urban spaces. These include the street patterns, road networks, building layouts, and the distribution of different land uses within the city. Street patterns, such as grid, radial, or organic layouts, influence the flow of traffic, accessibility, and connectivity between various parts of the city (Fig. 6.2). The density and height of buildings define the vertical and horizontal structure of urban spaces, contributing to the city's skyline, functionality, and character. Urban edges, such as boundaries, waterfronts, or walls, also play a key role in demarcating different areas of the city and regulating interaction between different zones. The arrangement and design of public spaces, like squares, parks, and recreational areas, further complement the structural layout, facilitating social interactions and community building. Together, these structural elements determine how the city functions, how people navigate it, and how different areas are connected, ensuring efficient land use and sustainable urban growth.

Fig. 6.2 Sample view of structural elements in urban centre

6.2.3 *Social and Cultural Elements*

Social and cultural elements of urban morphology encompass the ways in which cities reflect and shape the social dynamics, values, and identities of their inhabitants. Social elements include the spatial organization of residential, commercial, and public areas, which influence how communities interact, form networks, and use urban spaces. The distribution of amenities, such as schools, hospitals, markets, and recreational areas, reflects the city's prioritization of public welfare and social cohesion (Fig. 6.3). Cultural elements, on the other hand, are expressed through architecture, public monuments, heritage sites, and the overall design of urban spaces, which embody the historical, artistic, and cultural identities of a city. These cultural features often serve as markers of a community's collective memory, traditions, and values, providing a sense of place and belonging. The integration of social and cultural elements in urban design promotes inclusivity, fosters community engagement, and enhances the urban environment's ability to meet the diverse needs of its residents. Together, these elements ensure that cities are not only functional but also vibrant, reflective of the social and cultural fabric of their populations.

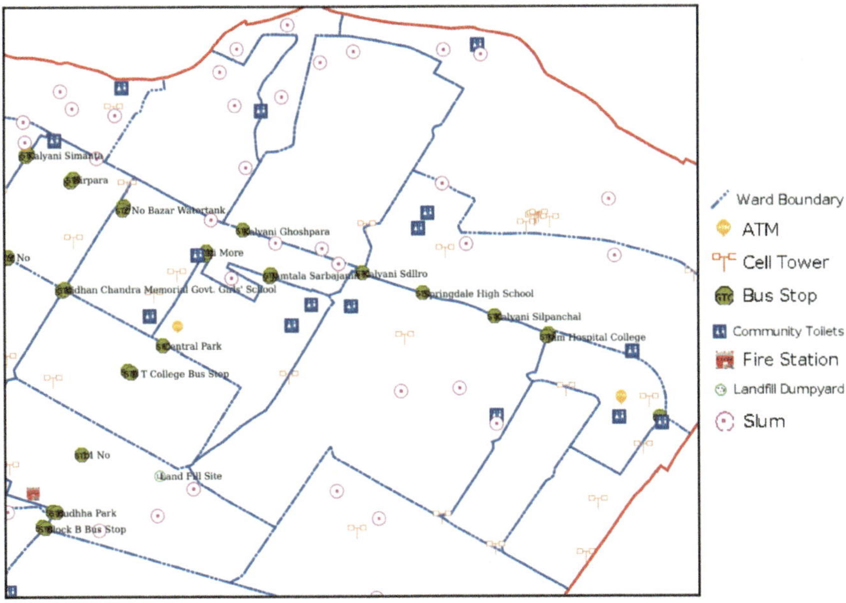

Fig. 6.3 Sample view of social and cultural elements in urban centre

6.2.4 Natural and Environmental Elements

Natural and environmental elements of urban morphology refer to the influence of the natural landscape and ecological features on the development and structure of urban areas. These elements include the topography, climate, hydrology, vegetation, and natural resources that shape the physical and functional characteristics of a city (Fig. 6.4). The natural landscape, such as hills, rivers, coastlines, and valleys, determines how the city expands, the location of buildings, and the use of land for agriculture, industry, or recreation. Climate plays a key role in shaping architectural designs and urban planning, influencing the positioning of green spaces, water management systems, and energy-efficient buildings. Hydrological factors, including rivers, lakes, and water tables, impact urban drainage systems, flood control measures, and the placement of water-based amenities. Urban greenery, parks, and open spaces are crucial for environmental sustainability, offering recreational areas, improving air quality, and promoting biodiversity. By integrating these natural elements into urban planning, cities can create more sustainable, resilient, and livable environments that harmonize with the surrounding ecosystem.

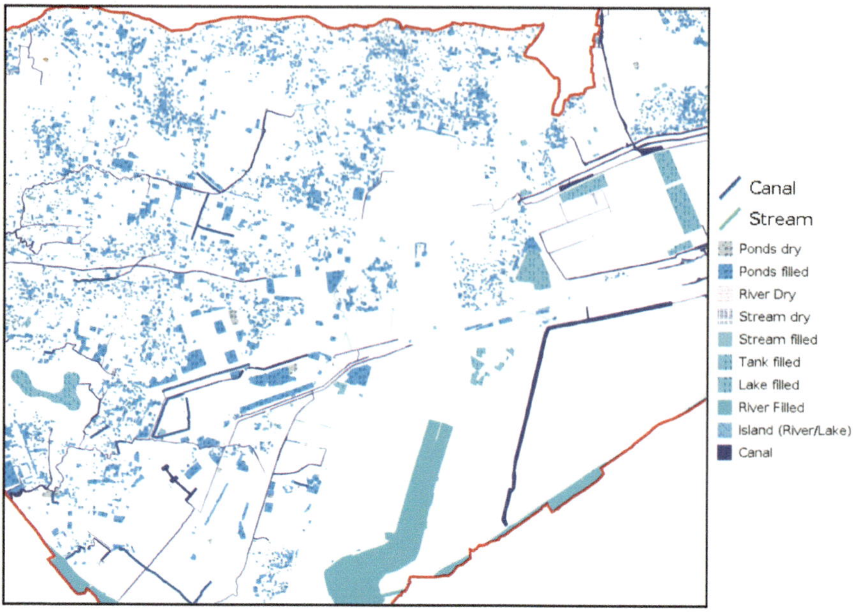

Fig. 6.4 Sample view of environmental elements in urban centre

6.3 Relationship Between Urban Form and Function

The relationship between urban form and function is fundamental in shaping how a city operates, grows, and impacts its residents. Urban form refers to the physical layout and spatial organization of the city, including its streets, buildings, public spaces, and infrastructure, while urban function pertains to how these physical elements are used to meet the needs and activities of the population, such as living, working, recreation, and transportation. The design and arrangement of urban spaces directly influence the efficiency and effectiveness of these functions. For example, a well-connected street network can facilitate easy movement and accessibility, supporting the economic function of the city by enabling the flow of goods and people. Similarly, the distribution of residential, commercial, and recreational areas impacts social interactions, livability, and the quality of life for citizens. Urban form, whether it's compact or sprawling, can also determine sustainability outcomes, such as energy consumption, environmental impact, and social equity. A well-planned urban form supports the diverse functions of the city, promoting a harmonious and efficient environment where all aspects of urban life can thrive.

6.4 Case Study—Distribution of Built-Up Area in Berhampore Town, West Bengal

Landscape statistics provide critical metrics for evaluating spatial patterns and understanding the ecological and developmental dynamics of an area. The geographical distribution of built-up areas in Berhampore highlights the need for sustainable urban planning. Berhampore, the district headquarters of Murshidabad in West Bengal, has experienced a dynamic transformation in its built-up area due to urbanization, population growth, and socio-economic development. The spatial distribution of built-up areas in Berhampore reflects both historical patterns of settlement and contemporary pressures of urban expansion.

Strategic measures, including zoning regulations, preservation of green spaces, and equitable infrastructure development, are essential to balance urban growth with environmental conservation and societal well-being. Landscape proportion measures the percentage of a specific land cover type within the total study area, highlighting the dominance or scarcity of features such as built-up regions, forests, or water bodies. Edge density, which quantifies the length of boundaries between different land cover types per unit area, reflects the complexity and fragmentation of the landscape, with higher values often indicating intensified human activity or habitat disruption. Patch density, measuring the number of distinct patches per unit area, assesses landscape heterogeneity, where a high density suggests fragmentation or diversified land use.

6.4.1 Methods

The Largest Patch Index (LPI) provides insight into spatial dominance, representing the proportion of the landscape occupied by the largest patch; this can signal landscape homogeneity or the prevalence of specific land use types. Meanwhile, the Patch Cohesion Index evaluates connectivity among patches of the same type, offering a measure of ecological linkage and habitat integrity—higher values suggest better physical continuity (Table 6.1). Lastly, the Splitting Index assesses landscape division into smaller, disconnected patches, with higher values indicating greater fragmentation and potential ecological challenges, such as reduced habitat connectivity.

Table 6.1 Estimation of landscape matrix

Year	Landscape proportion	Edge density	Patch density	Largest patch index	Patch cohesion index	Splitting index
2000	0.93408	0.08842	0.00308	91.67703	9.85392	1.18969
2023	0.90289	0.04516	0.00122	88.34694	9.95733	1.28117
2010	0.7761	0.20162	0.00758	71.70774	9.87181	1.94408

6.4.2 Results and Discussion

The central parts of Berhampore, particularly around the administrative and commercial hubs, exhibit the highest concentration of built-up areas (Fig. 6.5). These regions are characterized by dense residential zones interspersed with commercial establishments, government offices, and educational institutions. The built-up intensity here is high, with limited open spaces, reflecting a traditional urban fabric that has evolved over decades.

In recent years, significant expansion of built-up areas has been observed in the peripheries of Berhampore. These regions, which were once dominated by agricultural fields and rural settlements, are now transforming into suburban zones with the development of housing complexes, small industries, and infrastructure. Specific zones in Berhampore, such as areas near the Berhampore railway station and surrounding markets, have seen an increase in commercial and industrial activities.

The temporal analysis of landscape metrics for the years 2000, 2010, and 2023 reveals significant changes in the spatial structure and composition of the study area. Landscape Proportion, a measure of the dominance of specific land cover types, shows a gradual decline from 0.93408 in 2000 to 0.90289 in 2023, with the most notable drop occurring by 2010 (0.7761). This suggests a reduction in the extent of dominant land cover types, likely driven by urbanization or land conversion.

Edge density, reflecting boundary complexity, decreased significantly from 0.08842 in 2000 to 0.04516 in 2023, with an anomalous spike to 0.20162 in 2010. The increased edge density in 2010 may indicate heightened fragmentation or increased land use heterogeneity during this period. Patch density, representing the number of patches per unit area, mirrors this trend, peaking in 2010 at 0.00758, before declining to 0.00122 in 2023. This indicates an initial increase in landscape fragmentation and patchiness, followed by a consolidation of patches over time.

The Largest Patch Index (LPI), which indicates the dominance of the largest land cover patch, also declined from 91.67703 in 2000 to 88.34694 in 2023, reflecting a reduction in spatial dominance and increased diversity of smaller land cover types. Conversely, the Patch Cohesion Index improved marginally from 9.85392 in 2000 to 9.95733 in 2023, implying enhanced connectivity among patches, possibly due to natural restoration or strategic planning efforts.

Finally, the Splitting Index, which measures the degree of fragmentation, increased steadily from 1.18969 in 2000 to 1.28117 in 2023, with a sharp rise in 2010 (1.94408). This growth indicates progressive fragmentation over the years, potentially driven by urban sprawl and infrastructure development.

The landscape diversity of Berhampore town, as assessed through key diversity indices, reveals significant temporal changes in spatial and ecological heterogeneity between 2000 and 2023. The Shannon Index, a measure of landscape diversity that accounts for both richness and evenness of land use types, showed an increase from 0.24294 in 2000 to a peak of 0.531794 in 2010, followed by a decline to 0.31867 in 2023. This trend indicates that landscape diversity grew substantially during the early

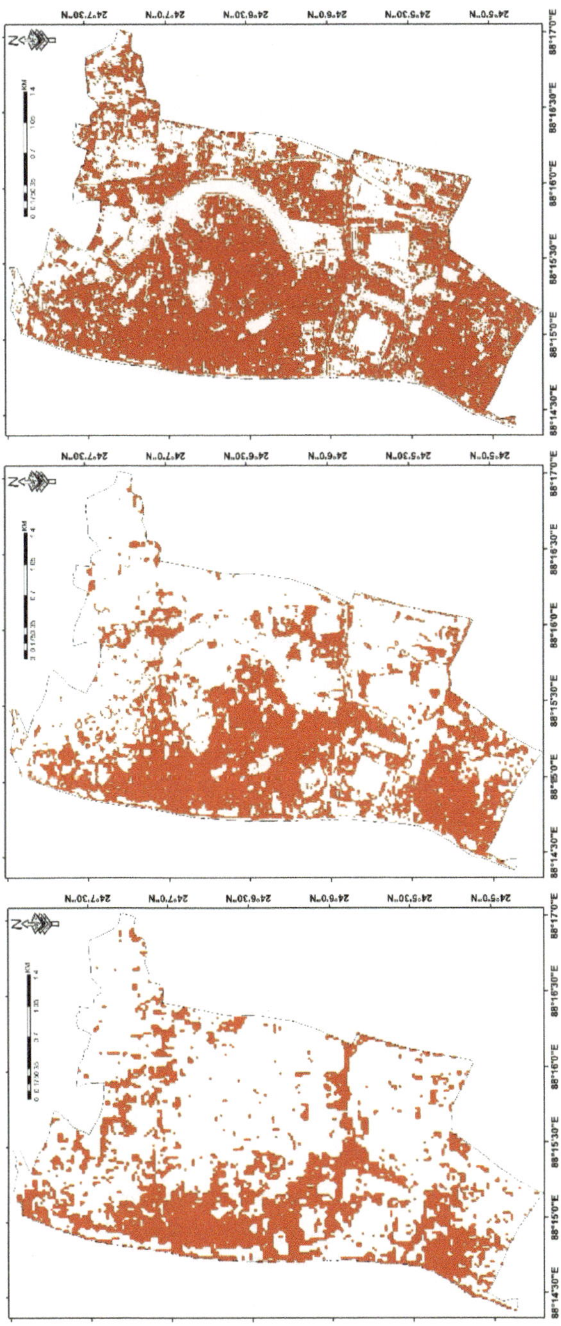

Fig. 6.5 Geographical distribution of built-up area in Berhampore town of West Bengal

Table 6.2 Landscape diversity matrix in the study area

Landscape index	2000	2010	2023
Shannon index	0.24294	0.531794	0.31867
Shannon equitability	0.35048	0.767217	0.45974
Simpson index	0.12314	0.347532	0.17535

2000s, likely due to the emergence of new land use types and increased heterogeneity (Table 6.2). However, the subsequent decline suggests a consolidation of specific land uses or a reduction in the variability of land cover types over time.

Similarly, the Shannon Equitability, which measures the evenness of distribution among different land use types, exhibited a pronounced increase from 0.35048 in 2000 to 0.767217 in 2010, followed by a reduction to 0.45974 in 2023. The peak in 2010 implies a more balanced distribution of various land cover types during this period, whereas the drop in 2023 suggests a dominance of certain land uses, potentially driven by urban sprawl and infrastructural expansion.

The Simpson Index, another indicator of diversity that emphasizes the probability of two randomly selected areas belonging to the same land use type, increased from 0.12314 in 2000 to 0.347532 in 2010, before decreasing to 0.17535 in 2023. The peak in 2010 reflects heightened landscape heterogeneity and the coexistence of diverse land use types, while the decline by 2023 aligns with the homogenization of certain land uses.

These shifts in landscape metrics underline dynamic land use and land cover transformations in the study area, with significant implications for ecological stability, habitat connectivity, and sustainable urban planning. Understanding these temporal changes is crucial for formulating adaptive land management and conservation strategies. The landscape diversity indices collectively portray a dynamic yet uneven trajectory of landscape diversity in Berhampore. The increase in diversity and equitability during the 2000s points to a phase of transformation and adaptation, whereas the declines in these metrics in recent years indicate challenges related to urban consolidation, land use homogenization, and potential ecological stress. Understanding these shifts is critical for framing policies that promote sustainable urban planning, biodiversity conservation, and balanced socio-economic development in Berhampore.

6.5 Tools and Techniques for Analysing Urban Morphology

Tools and techniques for analysing urban morphology involve a combination of spatial, quantitative, and qualitative methods to understand the structure and dynamics of urban environments. Geographic Information Systems (GIS) are central in mapping and analysing the spatial configuration of cities, enabling the study of land use patterns, street networks, and the relationship between built and natural environments. Remote sensing technologies, such as satellite imagery and aerial photography, allow for the observation of large-scale urban forms and their changes over time. Additionally, space syntax analysis is used to assess the connectivity and accessibility of urban spaces, providing insights into how people move through cities and how spaces interact. Urban simulation models help forecast the impacts of different planning strategies, considering factors like population growth, transportation, and environmental changes. Historical and archival research techniques can also be applied to understand the evolution of urban forms and their cultural significance. By integrating these tools, planners, researchers, and urban designers can gain a comprehensive understanding of urban morphology and make informed decisions about sustainable and functional urban development.

6.6 Interplay Between Urban Morphology and Smart City Components

The interplay between urban morphology and smart city components is crucial for creating sustainable, efficient, and livable urban environments. Urban morphology, which focuses on the physical form and spatial organization of cities, must be harmonized with the technologies and infrastructure that define a smart city. Smart city components, such as Internet of Things (IoT) sensors, data analytics, smart transportation systems, and energy-efficient infrastructure, can optimize urban functions, making cities more responsive to the needs of their residents. For instance, the layout of streets and public spaces in urban morphology can be enhanced by smart transportation systems that monitor traffic and provide real-time data, improving mobility and reducing congestion. Similarly, the integration of green spaces and sustainable architecture in urban design can be complemented by smart energy grids and waste management systems that reduce environmental impact and enhance resource efficiency. The efficient use of space, connectivity, and accessibility in urban morphology, combined with smart technologies, creates a dynamic and adaptive city where infrastructure responds proactively to both human behaviour and environmental factors, improving the quality of life for its inhabitants.

6.7 Spatial Planning and Urban Design

Spatial planning and urban morphology are closely linked in shaping the development, design, and functionality of cities. Spatial planning refers to the process of organizing land use, infrastructure, and urban development to meet the needs of a growing population, while urban morphology focuses on the physical form and layout of the city—how buildings, streets, open spaces, and other elements are arranged and interact. Key characteristics of urban morphology include the city's street patterns (grid, radial, or organic), the density and arrangement of buildings, the hierarchy of spaces (e.g., public vs. private spaces), and the integration of natural features such as parks, rivers, and green corridors. These elements define the spatial structure of a city, influencing how people move, interact, and experience urban life.

In terms of design, spatial planning uses these morphological characteristics to optimize land use and create sustainable, livable environments. For example, dense, mixed-use areas may promote walkability and reduce reliance on cars, while green spaces can improve air quality and provide recreational opportunities. The planning process also considers zoning regulations, transportation networks, housing, public services, and environmental sustainability. Effective urban design takes into account both functional needs (such as accessibility, safety, and connectivity) and aesthetic values, creating a balanced and cohesive urban environment.

6.8 Role of Density, Diversity, and Connectivity

Density, diversity, and connectivity are key elements of urban morphology that play a significant role in shaping the functionality, sustainability, and livability of cities. These three aspects interact with each other to influence how urban spaces are organized, how people move through them, and how effectively the urban environment meets the needs of its residents.

The density of a city refers to the concentration of buildings, people, or activities within a given area. High-density urban areas often promote efficient land use, as they maximize space while offering proximity to jobs, services, and amenities. Dense urban environments can foster vibrant communities, increase the accessibility of resources, and support public transport systems, reducing reliance on private cars. However, excessive density without proper infrastructure and green spaces can lead to overcrowding, environmental degradation, and reduced quality of life. Proper management of density is crucial to balancing the benefits of urbanization with the needs for open spaces, environmental sustainability, and liveability.

Urban diversity refers to the variety of land uses, building types, and activities within a city. A diverse urban form supports mixed-use development, where residential, commercial, and recreational spaces coexist. This diversity encourages social interaction, fosters economic activity, and allows for greater adaptability in the face of changes in population or economic conditions. Diversity in urban morphology also refers to the range of building styles, architectural heritage, and cultural expressions that give a city its unique character. The more diverse a city's-built environment and activities, the more resilient and vibrant it becomes, offering a wide range of opportunities for its inhabitants.

Connectivity in urban morphology is the degree to which different parts of the city are linked together, both physically and functionally. This includes the connectivity of transportation networks (e.g., roads, public transport, cycling paths) and the integration of different urban functions (e.g., commercial, residential, industrial, and recreational spaces). High connectivity improves accessibility, reduces travel time, and enhances social interaction, making cities more efficient and livable. A well-connected city also ensures the seamless flow of goods, services, and information, boosting economic activity and fostering social inclusion. Effective connectivity is especially important in dense, diverse urban areas, where it can mitigate potential congestion and ensure equal access to resources and opportunities.

6.9 Influence of Urban Morphology on Smart Infrastructure

Urban morphology significantly influences the development and functionality of smart infrastructure in cities. Smart infrastructure refers to the integration of advanced technologies—such as sensors, data analytics, and automation systems—into urban systems to improve their efficiency, sustainability, and responsiveness. The physical layout and structural characteristics of a city, as determined by urban morphology, play a key role in how these technologies are implemented, interact with the environment, and impact the daily lives of residents.

6.9.1 *Impact of Density on Smart Infrastructure*

High-density urban areas, with their concentrated population and building activity, are more likely to benefit from the economies of scale provided by smart infrastructure. In such environments, technologies like smart grids, waste management systems, and efficient public transport networks can be integrated and optimized more easily. For example, sensors and data analytics in high-density areas can help in managing energy consumption, reducing traffic congestion, or monitoring air quality. However, high-density areas may also face challenges such as overcrowded infrastructure and limited space for new installations, which requires careful planning of smart infrastructure integration.

6.9.2 *Density Analysis*

Garbage collection points within Bankura Municipality were meticulously identified and documented through a comprehensive field survey. This survey provided detailed spatial data, which was then incorporated into a Geographic Information System (GIS) software to create a structured spatial database. The database serves as the foundation for analysing the distribution and density of garbage collection points across the municipality.

Using GIS tools, specifically the point density analysis feature in QGIS software, the distribution pattern of garbage collection points was studied. Point density analysis is a geospatial technique that calculates the concentration of specific points (in

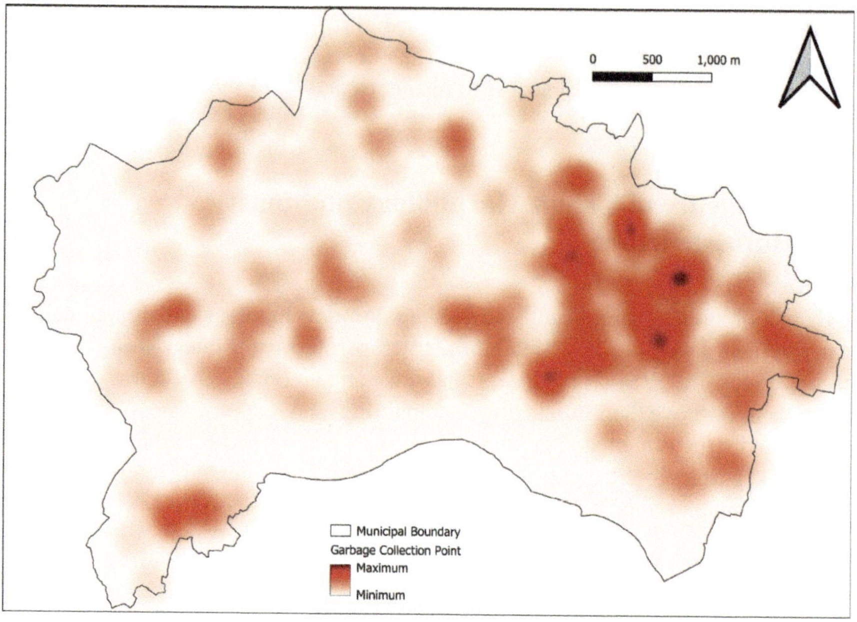

Fig. 6.6 Spatial variation of garbage collection point

this case, garbage collection points) within a defined area (Fig. 6.6). Each garbage collection point represents a count or quantity that contributes to the creation of a continuous surface. This continuous surface visually represents areas with varying densities of garbage collection points, highlighting zones with higher or lower concentrations.

The output of this analysis is a raster dataset, where each cell represents a calculated density value. The cell size for the raster dataset was carefully chosen to balance resolution and computational efficiency. A smaller cell size provides higher detail but requires more computational resources, while a larger cell size simplifies the dataset at the expense of finer spatial granularity. The resulting raster dataset provides a critical input for urban planning and decision-making, helping identify areas requiring additional garbage collection points or better resource allocation.

6.9.3 K-Means Clustering

K-means clustering is a statistical and spatial analytical technique used to partition a set of data points into distinct groups or clusters. In the context of garbage collection points within a municipality, this method is applied to calculate the 2D distance-based cluster numbers for each point. By employing *K-means* clustering, the spatial distribution of garbage collection points is analysed and organized into *k* clusters,

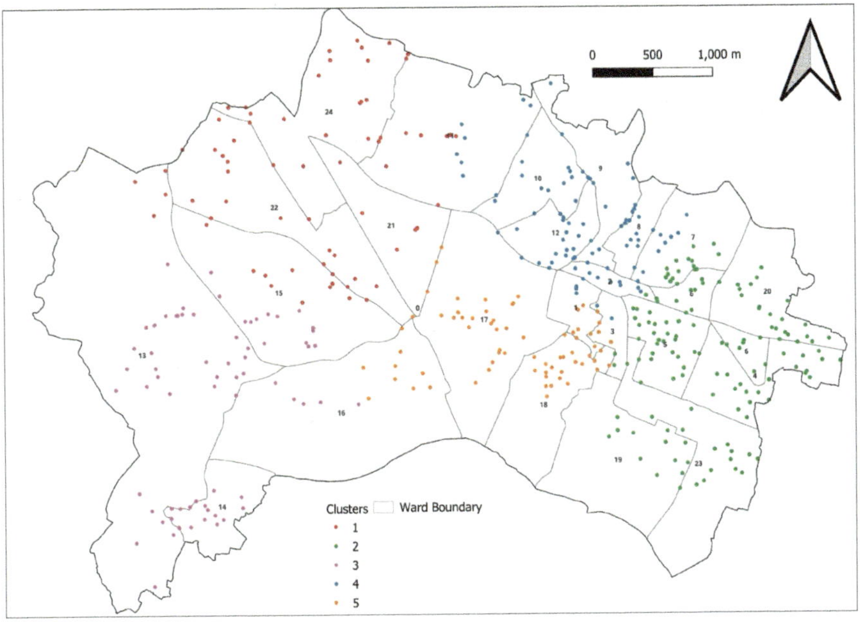

Fig. 6.7 Spatial cluster of garbage location point

where k is a user-defined parameter representing the number of clusters to be created (Fig. 6.7).

Methods of K-Means Clustering

- The algorithm begins by initializing k centroids randomly within the spatial extent of the garbage collection points. These centroids serve as initial guesses for the barycenters of the clusters.
- Each garbage collection point is assigned to the cluster represented by the nearest centroid. The proximity is calculated based on the 2D Euclidean distance between the point and each centroid. This assignment minimizes the within-cluster variance, ensuring that each point is grouped with others located spatially closer to it.
- Once all points have been assigned to clusters, the algorithm recalculates the position of each centroid. The new centroid of a cluster is computed as the mean (barycenter) of all garbage collection points assigned to that cluster. This process updates the centroids to better represent the spatial centre of their respective clusters.
- Steps 2 and 3 are repeated iteratively. In each iteration, points may be reassigned to different clusters based on the updated centroids. The process continues until the centroids stabilize (i.e., no significant change in position) or a predefined number of iterations is reached.

- The final output assigns each garbage collection point a cluster number. This number indicates the cluster to which the point belongs. The clustering result represents spatial groupings of garbage collection points with minimal intracluster distance and maximal intercluster separation.

The clusters generated through *K-means* clustering are valuable for urban and waste management planning. By identifying spatial groupings of garbage collection points, municipalities can:

- Optimize the allocation of waste collection resources.
- Plan efficient routes for garbage collection vehicles.
- Identify areas with sparse or dense distribution of collection points for better infrastructure planning.
- The choice of *k (number of clusters)* significantly influences the clustering results. Selecting an appropriate *k* is critical and may require experimentation or the use of methods such as the elbow method.
- *K-means* clustering assumes that clusters are roughly spherical and equally sized, which may not always reflect real-world spatial patterns.

6.10 Role of Connectivity in Smart Infrastructure

Connectivity, both in terms of physical infrastructure (such as roads, public transit, and utilities) and digital networks, is essential for the functioning of smart systems. In cities with well-connected urban morphology, smart infrastructure can be effectively coordinated, ensuring that various systems (e.g., transportation, water supply, waste management) are interlinked and can communicate in real time (Fig. 6.8). Connectivity enhances the ability of technologies like smart traffic lights, IoT-based environmental monitoring, and adaptive energy systems to function smoothly across a city. Moreover, areas with good connectivity between digital and physical networks enable better data collection and management, which in turn improves the responsiveness and adaptability of smart infrastructure.

6.11 Influence of Diversity on Smart Infrastructure

The diversity of urban form, including mixed-use areas and varying building types, presents both opportunities and challenges for smart infrastructure. In diverse environments, the implementation of smart infrastructure must account for a wide range of user needs and urban functions. For instance, a smart city with a mix of residential, commercial, and industrial spaces will require smart solutions tailored to each function, such as building automation systems for residential areas, energy-efficient designs for commercial spaces, and smart waste management solutions for industrial zones. Moreover, diverse neighbourhoods may have different levels of technological

Fig. 6.8 Sample view of smart infrastructure

readiness, necessitating targeted interventions to ensure equitable access to smart infrastructure and services.

6.12 Adaptability to Urban Form and Environmental Challenges

Urban morphology also affects the adaptability of smart infrastructure to environmental and climate challenges. For example, cities with varied topography or those located in areas prone to flooding or other natural hazards may require specialized smart infrastructure to address these risks. This could include flood monitoring systems, smart drainage, or resilient energy grids. Additionally, urban form influences how green spaces, natural landscapes, and environmental features (such as waterways or hills) are integrated into the city's infrastructure, supporting smart, sustainable solutions like smart irrigation or renewable energy harvesting from natural resources.

6.13 Spatial Distribution of Resources

Urban morphology also shapes the spatial distribution of resources such as water, energy, and waste. Smart infrastructure systems must be designed to optimize the distribution and use of these resources. For example, in areas where infrastructure is sprawling or unevenly distributed, smart grids or energy-efficient systems may need to be more robust to manage the dispersed energy needs of the population. In contrast, compact urban areas can support more centralized systems and benefit from tighter management of resources.

6.14 Case Studies: Urban Morphology in Smart City Projects

6.14.1 Analysing Spatial Inequity of Urban Services in Bengaluru

Spatial inequity in urban services is a critical issue in rapidly urbanizing cities like Bengaluru, where population growth and infrastructure development often fail to align. The uneven distribution of services such as healthcare, education, public transport, and sanitation can exacerbate social and economic disparities, disproportionately affecting marginalized communities. Incorporating global population data into this analysis provides a robust framework to understand these inequities and identify areas of intervention.

Bengaluru's dynamic growth, characterized by a mix of planned and unplanned development, creates significant challenges in the equitable distribution of urban services. Using high-resolution global population data, such as those derived from datasets like WorldPop or LandScan (Fig. 6.9), we can spatially map population densities across the city. This data, when combined with geospatial information on the availability and quality of services, enables a detailed analysis of service accessibility relative to population distribution. For instance, by overlaying the location of healthcare facilities with population density, we can identify underserved areas where residents must travel long distances for basic healthcare. Similarly, analysing public transport routes and stops in relation to population clusters can reveal gaps in connectivity, leaving certain neighbourhoods isolated. This approach can also extend to examining access to clean water, waste management, educational institutions, and green spaces.

In Bengaluru, certain wards and peri-urban areas may exhibit stark disparities compared to the well-serviced central zones. Migrant populations, informal settlements, and rapidly expanding suburbs are often the most affected. Using global population data enhances the precision of such analyses, offering a scalable and consistent approach to urban planning. This spatial inequity analysis can inform

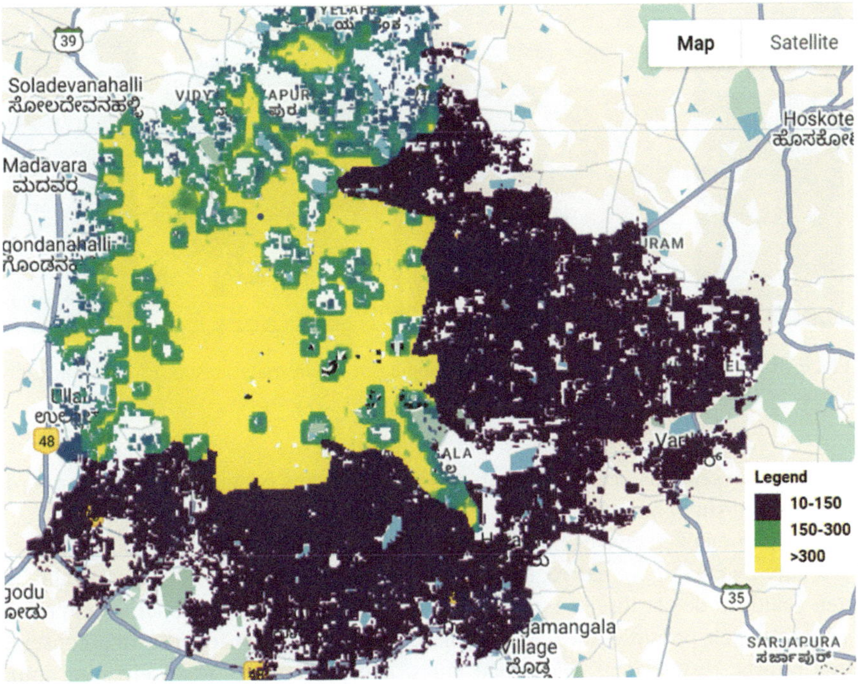

Fig. 6.9 Projected population of urban services

policymakers and planners in prioritizing investments and reforms. For example, it can guide decisions on locating new facilities, optimizing transport routes, and improving basic service delivery in underserved regions. Moreover, such data-driven insights can play a pivotal role in fostering inclusive urban development, aligning with Bengaluru's vision of becoming a smart, sustainable, and equitable city.

6.14.2 Mapping Biodiversity Hotspots Through a Geo-enabled Tree Census in Pune

The Pune Municipal Corporation (PMC) has embarked on an innovative geo-enabled tree census survey to comprehensively map biodiversity hotspots within its jurisdiction. This forward-thinking initiative aims to document the city's urban green cover and gain a holistic understanding of its biodiversity. By utilizing advanced geospatial technologies such as Geographic Information Systems (GIS) and Global Positioning System (GPS), the survey captures precise data on individual trees, including their species, age, height, girth, health, and geographic location. This data is then analysed to identify regions with high ecological value, referred to as biodiversity hotspots,

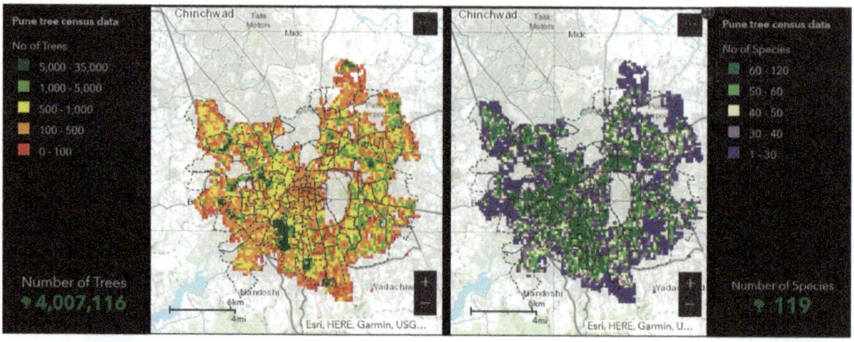

Fig. 6.10 Sample representation of spatial mapping of biodiversity hotspot (*Source* https://iuo.mohua.gov.in/portal/apps/sites/#/data)

which often serve as critical habitats for various forms of wildlife, including birds, insects, and small mammals.

The geo-enabled census not only records the tree population but also assesses its diversity by cataloging native, exotic, and endangered species, as well as evaluating tree health to detect threats such as diseases or pest infestations. This comprehensive approach provides the PMC with actionable insights to prioritize conservation efforts, restore degraded urban areas, and plan future urban greening initiatives. Furthermore, the spatial mapping of biodiversity hotspots aids in urban planning, ensuring that development projects align with ecological sustainability goals. The initiative also contributes to building climate resilience by mitigating urban heat islands, enhancing carbon sequestration, and supporting natural ecosystems in adapting to environmental changes (Fig. 6.10).

By integrating cutting-edge technology with biodiversity management, PMC's geo-enabled tree census sets a benchmark in urban environmental conservation. It not only highlights the city's commitment to sustainable growth but also fosters community engagement by raising awareness about the importance of preserving green spaces. Through this initiative, Pune is charting a path toward becoming a smart, inclusive, and ecologically balanced city.

6.14.3 Bus Service Levels in Delhi—A Comprehensive Analysis

The bus service levels in Delhi play a crucial role in defining the accessibility and efficiency of the city's public transport network. A detailed analysis involves examining various factors, including the distribution of bus stops, the frequency of buses, and their spatial locations across the city. By mapping the shifting distribution of bus services across different wards throughout the day, we gain insights into the temporal and spatial variability of service levels.

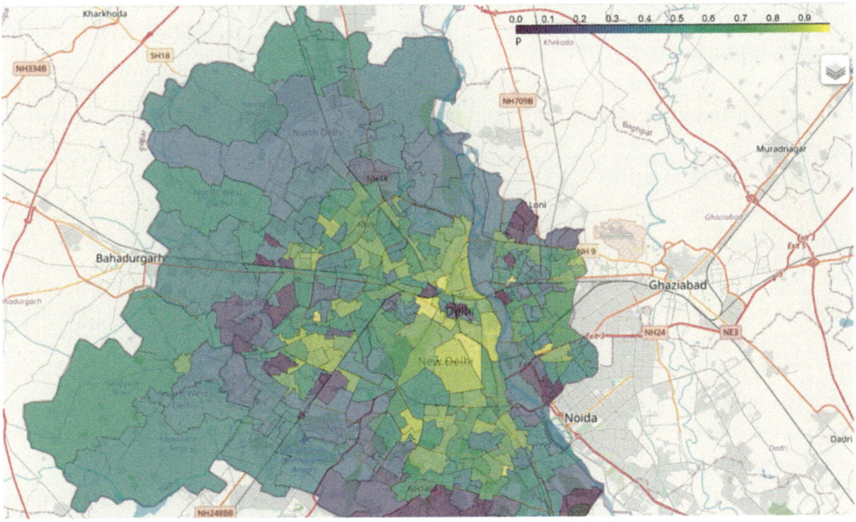

Fig. 6.11 Yellow means a higher proportion of population lives within 400 m of a bus stop, Purple means whereas Purple represents areas where a lower proportion of the population falls within this distance (*Source* https://iuo.mohua.gov.in/portal/apps/sites/#/data)

The distribution of bus stops is a key indicator of accessibility, as it determines the ease with which residents can reach public transport services. Similarly, the frequency of buses at these stops reflects the reliability and capacity of the system to cater to the population's mobility needs. Spatial mapping of these factors reveals patterns of connectivity, highlighting areas with robust service and identifying wards that might face inadequate coverage (Fig. 6.11).

By applying this analysis uniformly across all wards, it is possible to calculate the proportion of the population in each ward that resides within walking distance of a bus stop. This metric provides a direct measure of accessibility and identifies disparities in service provision. The outcome of such an analysis can guide urban planners and policymakers in prioritizing interventions, such as adding bus stops, increasing service frequencies, or redistributing buses to underserved areas, thereby ensuring equitable access to public transport.

6.15 Technological Integration and Urban Morphology

The interplay between technological advancements and urban morphology—the study of the physical form and structure of cities—has profoundly shaped the evolution of human settlements. With the rapid progression of digital, computational, and data-driven technologies, cities have become complex ecosystems where infrastructure, human activity, and spatial organization converge dynamically. When integrated, technology transforms traditional urban morphology by offering innovative methods for mapping, modelling, and managing urban spaces. Smart city initiatives exemplify this convergence, where technology is leveraged to optimize transportation networks, improve utility services, and foster community engagement. The fusion of 3D modelling, digital twins, and big data analytics has also redefined how urban planners visualize and implement growth strategies. Moudon (1997) emphasized the role of urban form analysis in understanding city development, while Conzen (1960) pioneered the concept of morphological regions, which underpin studies on urban transformation. These frameworks provide the baseline for integrating technology into urban studies. Batty (2005) highlighted how GIS facilitates spatial data visualization, enabling a granular analysis of urban forms. Additionally, 3D modelling and digital twin technologies, as discussed by Du et al. (2020), offer dynamic simulations of urban growth, enhancing predictive urban planning.

However, technological integration in urban morphology presents challenges, including ensuring equitable access, addressing privacy concerns, and preserving cultural identity amidst rapid modernization. Striking a balance between leveraging technology and maintaining human-centric, inclusive urban design is essential for sustainable urban futures. This intersection marks a pivotal era in city planning and management, where urban morphology serves as a foundation for understanding the complexities of space, while technological integration provides tools to innovate and evolve cities to meet modern demands. The integration of IoT sensors in urban design, as detailed by Albino et al. (2015), creates adaptive systems that redefine the relationship between infrastructure and urban form. As Graham and Marvin (2001) argued, the "splintering urbanism" caused by uneven technological adoption risks exacerbating socio-spatial inequalities. Recent studies focus on using AI and machine learning to enhance urban morphological analysis. For example, applications in land use classification (Chen et al. 2020) and urban heat island mapping demonstrate the potential of technology to inform sustainable city planning.

6.16 GIS and Spatial Data Analytics in Urban Morphology

Urban morphology examines the physical form and structure of urban areas, encompassing elements such as street patterns, building density, land use, and the spatial organization of cities. In recent years, Geographic Information Systems (GIS) and spatial data analytics have become indispensable tools for analysing and interpreting these components with unprecedented precision and depth. Early studies focused on mapping urban growth (Batty and Longley 1994) and understanding the spatial arrangement of buildings and streets. Techniques such as spatial autocorrelation (Anselin 1995), network analysis, and clustering algorithms have been instrumental in studying urban density, connectivity, and land use patterns. Advances in big data analytics have further enriched these studies by incorporating real-time data from sensors, satellites, and social media (Arribas-Bel 2014).

GIS enables the integration, visualization, and analysis of diverse spatial datasets, allowing urban planners, architects, and geographers to understand the dynamics of city growth and structure. Through spatial data analytics, patterns and trends in urban development can be identified, facilitating informed decision-making. Techniques such as heatmaps, network analysis, and 3D modelling provide detailed insights into urban density, accessibility, and the relationship between built environments and natural systems. For instance, studies have used GIS to analyse land use changes and their environmental impacts (Herold et al. 2003) and to optimize urban transportation networks (Zhang and Kukadia 2005). Predictive models and simulations are increasingly used to forecast urban growth and assess the implications of planning decisions. The use of 3D GIS and visualization tools has enhanced the understanding of vertical urban growth and its effects on spatial interactions (Biljecki et al. 2015).

Key applications include urban planning, where GIS aids in zoning and resource allocation; transportation, where network analytics optimize routes and connectivity; and sustainability, where spatial analytics assess the impact of urbanization on ecosystems and propose mitigation strategies. Additionally, emerging technologies like machine learning and AI in spatial analytics are enabling predictive modelling, offering valuable foresight into future urban growth scenarios. As cities face challenges like rapid urbanization, climate change, and resource scarcity, GIS and spatial data analytics stand at the forefront of creating smarter, more resilient urban environments. This fusion of technology and urban morphology not only provides actionable insights but also supports the development of sustainable and inclusive cities.

6.17 IoT Applications in Urban Infrastructure

The Internet of Things (IoT) is revolutionizing urban infrastructure by integrating smart technologies into the management of cities. By connecting devices, sensors, and systems, IoT enables real-time data collection, analysis, and communication, paving the way for efficient and sustainable urban environments. Studies highlight the use of IoT for smart traffic lights, vehicle-to-infrastructure communication, and congestion monitoring (Zanella et al. 2014). Smart meters for electricity and water consumption enable precise monitoring and billing, while leak detection sensors and automated systems optimize resource use (Gungor et al. 2013).

In transportation, IoT improves traffic management through smart traffic lights, real-time tracking of public transit, and connected vehicles. For utilities, smart grids and IoT-enabled water systems enhance energy efficiency and resource allocation. Waste management benefits from IoT-enabled bins that monitor fill levels and optimize collection routes. Similarly, environmental monitoring uses IoT sensors to track air quality, noise levels, and weather conditions, aiding in urban sustainability efforts. Real-time air pollution monitoring systems, for example, aid in addressing urban air quality issues (Dutta et al. 2020).

Smart buildings and infrastructure rely on IoT to optimize energy use, monitor structural health, and ensure occupant safety. Public safety systems leverage IoT through connected surveillance cameras and emergency response systems. Additionally, IoT facilitates citizen engagement via apps that provide real-time updates on city services and enable feedback loops. Smart building systems use IoT devices to regulate HVAC systems, detect maintenance needs, and ensure safety through connected alarms and surveillance (Khajenasiri et al. 2017). In infrastructure, IoT helps monitor the health of bridges, roads, and other structures, predicting failures and ensuring timely repairs. Studies emphasize the role of IoT in connecting first responders and streamlining disaster management workflows (Perera et al. 2014).

As cities grow, IoT applications in urban infrastructure offer transformative potential to address challenges such as congestion, pollution, and resource scarcity. By fostering innovation and collaboration, IoT serves as a cornerstone for building smart, sustainable, and resilient cities of the future.

6.18 Role of AI and Machine Learning in Urban Pattern Analysis

Artificial Intelligence (AI) and machine learning (ML) are transforming urban pattern analysis by providing innovative tools to identify, predict, and optimize spatial and temporal trends in urban development. These technologies enable the analysis of vast datasets, uncovering complex relationships and insights that traditional methods might overlook.

AI and ML enhance the study of urban morphology by automating the detection of land use changes, mapping urban sprawl, and analysing transportation networks. Predictive models, powered by ML algorithms, forecast future urban growth and its implications for infrastructure, housing, and environmental sustainability. Deep learning methods, such as convolutional neural networks (CNNs), are widely applied to process satellite imagery and geospatial data, enabling high-resolution urban classification and segmentation. Maxwell et al. (2018) emphasize the efficiency of deep learning models in extracting features from satellite imagery for urban classification. Liu et al. (2019) showcase the use of Cellular Automata (CA) combined with ML to simulate urban expansion scenarios.

Additionally, AI-driven tools support real-time analysis of urban dynamics, such as traffic flows, energy consumption, and population density, aiding in the optimization of urban resources. Reinforcement learning algorithms further enhance urban planning by simulating scenarios and evaluating the outcomes of policy decisions. Feng et al. (2024) developed an ML-based framework to analyse human mobility patterns using GPS and social media data. Zhou and Li (2025) applied deep learning for monitoring urban heat island dynamics using multitemporal satellite data. Sun et al. (2022) used clustering algorithms to map socio-economic divisions in urban settings based on census and spatial data. Mittelstadt et al. (2016)

highlight the importance of ethical considerations in AI-driven decision-making for urban environments.

The integration of AI and ML in urban pattern analysis is pivotal for smart city development, offering data-driven insights to tackle challenges like climate change, resource scarcity, and rapid urbanization. These technologies empower planners, policymakers, and researchers to design resilient and sustainable urban systems tailored to future needs (Fig. 6.12).

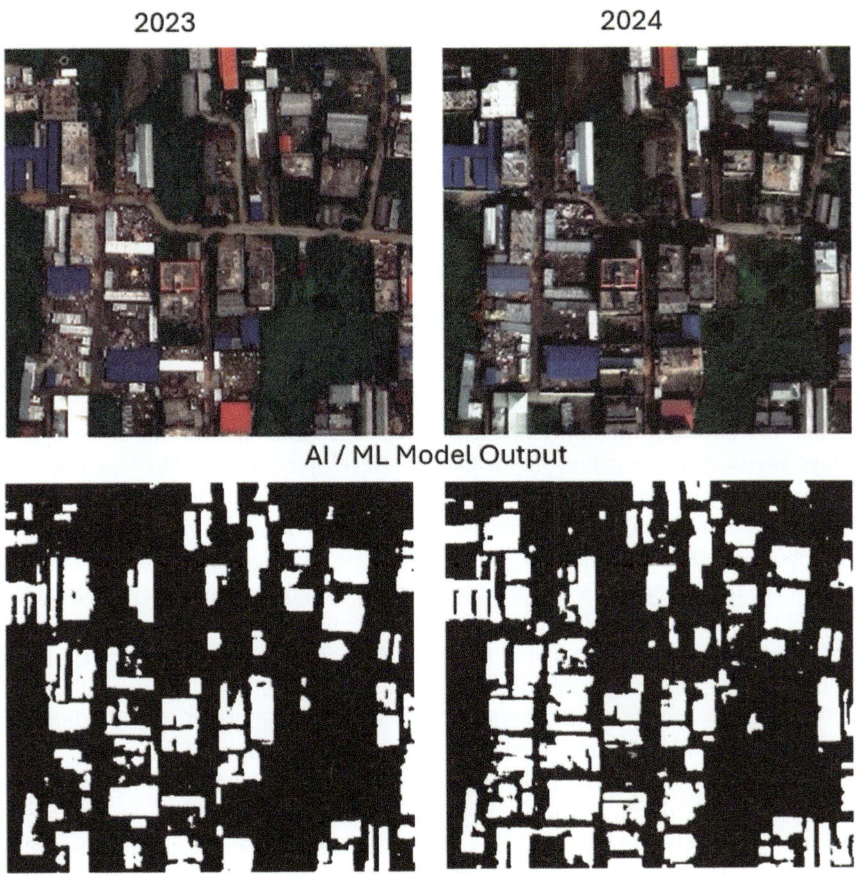

Fig. 6.12 AI/ML-based building footprint detection

6.19 Sustainability, Resilience, and Urban Morphology

"Sustainability, Resilience, and Urban Morphology" are interconnected concepts critical to the development of future urban landscapes. Sustainability focuses on creating urban environments that meet the needs of the present without compromising the ability of future generations to meet their own needs, particularly in terms of resource management, energy efficiency, and ecological preservation. Resilience refers to the ability of cities to adapt and recover from shocks such as natural disasters, climate change, or economic crises, ensuring continuity and long-term survival. Urban morphology, the study of the physical form of cities, plays a crucial role in shaping both sustainability and resilience. The design, structure, and layout of urban spaces influence factors like resource flow, accessibility, and vulnerability to risks. A well-designed urban morphology fosters sustainable development, enhances resilience to environmental pressures, and improves the overall quality of life for urban populations.

6.20 Urban Morphology and Sustainable Development Goals (SDGs)

Urban morphology, the study of the physical form and structure of urban spaces, plays a critical role in achieving Sustainable Development Goals (SDGs). Cities are dynamic entities that constantly evolve, shaped by complex interactions between natural systems, built environments, and societal needs. Understanding urban morphology offers insights into spatial patterns, infrastructure design, and land use, which are fundamental for sustainable urban development. The SDGs, particularly Goal 11 ("Make cities and human settlements inclusive, safe, resilient, and sustainable"), emphasize the importance of integrating environmental, social, and economic considerations into urban planning. By analysing urban form, policymakers and urban planners can identify strategies to reduce carbon footprints, optimize land use, enhance resilience to climate change, and improve the quality of urban life.

Studies emphasize that compact urban forms, characterized by higher density and mixed land use, promote sustainability by reducing transportation emissions and fostering walkability (Jenks and Jones 2010). Conversely, sprawled urban forms often result in inefficiencies, increased energy consumption, and habitat fragmentation. Compact cities with well-integrated green infrastructure are better equipped to withstand climate impacts, such as heatwaves and flooding (Gökçe 2020). Inclusive urban forms incorporate equitable distribution of amenities, efficient public transportation, and pedestrian-friendly designs, aligning with SDG targets for inclusive cities (Dovey and Pafka 2016). Case studies reveal that marginalized communities often reside in poorly connected areas, exacerbating inequalities. Sustainable urban morphology advocates for optimal land use, balancing development with conservation. Mixed-use developments and vertical urban growth are identified as strategies to minimize land consumption while accommodating growing populations (Neuman 2005). Advances in Geographic Information Systems (GIS) and remote sensing have revolutionized urban morphology studies. High-resolution spatial data enables precise mapping of urban structures, facilitating data-driven planning for SDGs (Batty 2013). Integrating AI and machine learning further enhances predictive modelling for sustainable urban transformations.

6.21 Case Study—Urban Morphology and SDGs in Kolkata Municipal Corporation

The survey reveals significant insights into the awareness and perceptions of urban morphology and its relationship with Sustainable Development Goals (SDGs) within the Kolkata Municipal Corporation (KMC). A majority of respondents (60%) are aware of the concept of urban morphology, though 40% are unfamiliar, highlighting the need for increased education and outreach. An overwhelming 70% of participants believe that urban morphology significantly influences SDGs, emphasizing its perceived importance in sustainable urban development. Among the SDGs, SDG 11 (Sustainable Cities and Communities) is identified by half of the respondents (50%) as the most impacted by urban morphology, followed by SDG 6 (Clean Water and Sanitation) at 30%, and SDG 13 (Climate Action) at 20% (Table 6.3).

Key challenges to achieving SDGs through urban morphology in Kolkata include unplanned growth (40%), lack of green spaces (30%), poor transportation networks (20%), and inefficient land use (10%). These findings point to critical areas for intervention. Despite these challenges, only 30% of respondents feel that urban planning in Kolkata is aligned with SDGs, while a significant 60% disagree, indicating gaps in policy and implementation. Stakeholder participation is also deemed inadequate by 60% of respondents, further underscoring the need for more inclusive planning processes.

Regarding solutions, 40% of respondents prioritize green infrastructure as essential for SDG alignment, followed by affordable housing (30%), resilient public transport (20%), and energy-efficient buildings (10%). A notable 80% believe that GIS-based tools can play a pivotal role in achieving SDG targets, demonstrating confidence in technological interventions. However, current policies addressing morphological issues are considered effective by only 20% of participants, with 70% expressing dissatisfaction. To align KMC's urban form with SDGs, better land use planning (40%), public–private partnerships (30%), community participation (20%), and technological innovations (10%) are identified as key strategies.

These findings highlight both challenges and opportunities for integrating urban morphology into sustainable urban development strategies in Kolkata, calling for coordinated efforts between stakeholders, policymakers, and technology providers.

Table 6.3 Survey on urban morphology and SDGs in Kolkata municipal corporation

Question	Response options	Number (*N*)	Percentage (%)
Are you aware of the concept of urban morphology?	Yes	180	60
	No	120	40
Do you think urban morphology influences sustainable development goals?	Yes	210	70
	No	90	30
Which SDG is most impacted by urban morphology in Kolkata?	SDG 11 (sustainable cities)	150	50
	SDG 6 (clean water and sanitation)	90	30
	SDG 13 (climate action)	60	20
What are the key morphological challenges in achieving SDGs in KMC?	Unplanned growth	120	40
	Lack of green spaces	90	30
	Poor transportation networks	60	20
	Inefficient land use	30	10
Do you think urban planning in Kolkata is aligned with SDGs?	Yes	90	30
	No	180	60
	Unsure	30	10
Which factors should be prioritized in KMC for SDG-aligned morphology?	Green infrastructure	120	40
	Affordable housing	90	30
	Resilient public transport	60	20
	Energy-efficient buildings	30	10
Is stakeholder participation adequate in planning for SDG alignment?	Yes	90	30
	No	180	60
	Unsure	30	10
Do you believe GIS-based tools can help achieve SDG targets in KMC?	Yes	240	80
	No	30	10
	Unsure	30	10
Are the current policies addressing morphological issues effectively?	Yes	60	20
	No	210	70

(continued)

Table 6.3 (continued)

Question	Response options	Number (*N*)	Percentage (%)
	Unsure	30	10
What is the most effective way to align KMC's urban form with SDGs?	Better land use planning	120	40
	Public–private partnerships	90	30
	Community participation	60	20
	Technological innovations	30	10

Source Primary survey

6.22 Designing Resilient Urban Spaces

The concept of resilient urban spaces is pivotal in addressing the challenges posed by rapid urbanization, climate change, and social inequities. Resilient urban design integrates adaptive, sustainable, and inclusive practices to create cities that can withstand and recover from disruptions while ensuring the well-being of their inhabitants. Key principles include robust infrastructure, green and blue networks, equitable access to resources, and participatory planning. As cities worldwide face increasing environmental and socio-economic pressures, designing resilient urban spaces emerges as a crucial strategy for sustainable development, aligning with global agendas like the Sustainable Development Goals (SDGs).

Resilient urban spaces are built on principles of adaptability, flexibility, and redundancy. Studies underscore the importance of multifunctional spaces that serve ecological, social, and economic purposes (Meerow et al. 2016). Addressing climate change requires embedding resilience into urban form. Research highlights the role of urban design in mitigating heat island effects, managing stormwater, and reducing carbon footprints (Chelleri et al. 2015). Case studies from Latin America reveal how community-led initiatives have transformed vulnerable neighbourhoods into resilient spaces (Pelling 2011). Emerging technologies, including Geographic Information Systems (GIS), IoT, and AI, enable precise analysis and predictive modelling for resilience planning. Studies highlight the use of smart sensors for real-time monitoring of urban systems, aiding in disaster preparedness and adaptive management (Batty 2013). Research identifies public–private partnerships and innovative financing mechanisms as key to funding resilience projects. Examples include resilience bonds and green infrastructure investments (Zevenbergen et al. 2018).

6.23 Mitigating Urban Heat Islands Through Morphological Planning

Urban heat islands (UHIs), characterized by elevated temperatures in urban areas compared to their rural surroundings, pose significant challenges to human health, energy consumption, and environmental sustainability. Morphological planning, which focuses on the spatial configuration and physical characteristics of urban environments, offers innovative solutions to mitigate UHIs. By strategically designing urban layouts, integrating green and blue infrastructure, and optimizing building forms and materials, cities can reduce heat accumulation and enhance thermal comfort.

Research shows that compact urban layouts can exacerbate heat retention, while designs promoting ventilation corridors and optimized street canyons can reduce temperatures (Oke 1988). Integrating vegetation and water bodies into urban design is a proven strategy for UHI mitigation. Parks, green roofs, and urban wetlands lower surface and air temperatures through shading and evapotranspiration (Bowler et al. 2010). Reflective and permeable materials play a crucial role in reducing heat absorption. High-albedo surfaces and cool roofs can lower urban temperatures, as demonstrated in studies focusing on material science and urban energy balance (Akbari et al. 2001). Case studies from cities like Tokyo and New York highlight successful implementation of tree-planting initiatives to combat UHI effects (Taha 1997). Advances in remote sensing and GIS enable precise UHI mapping and assessment. These tools support morphological planning by identifying heat hotspots and evaluating the impact of mitigation measures (Voogt and Oke 2003).

6.24 Policy Implications and Governance

Policy implications and governance play a pivotal role in shaping the outcomes of societal development initiatives by setting the framework for decision-making, resource allocation, and accountability. Effective governance ensures that policies are designed to address public needs, align with sustainable development goals, and adapt to changing socio-economic and environmental contexts. The integration of evidence-based policymaking, participatory governance, and transparent mechanisms fosters inclusivity and equity in addressing critical issues such as climate resilience, resource management, and technological advancements. Furthermore, governance structures that prioritize adaptability and collaboration among stakeholders can enhance policy coherence and implementation effectiveness, ultimately driving transformative and sustainable impacts at local, national, and global levels.

6.25 Urban Planning Policies and Morphological Guidelines

Urban planning policies and morphological guidelines are essential tools for shaping sustainable and resilient urban environments. These frameworks guide the spatial organization, infrastructure development, and functional zoning of urban areas, ensuring compatibility with socio-economic objectives and environmental constraints. Policies often aim to balance urban growth with the preservation of natural resources, promote mixed-use developments, and enhance accessibility and livability. Morphological guidelines focus on the physical form and structure of urban spaces, addressing issues such as building density, street layouts, green spaces, and architectural harmony. Together, these approaches play a critical role in fostering cohesive urban landscapes, mitigating the effects of rapid urbanization, and promoting community well-being.

Research on urban planning policies underscores the need for integrated approaches that address economic, social, and environmental dimensions (Alexander et al. 1977). Studies highlight the effectiveness of participatory planning in ensuring stakeholder engagement and equitable resource distribution. Morphological studies, on the other hand, emphasize the importance of compact urban forms and the interplay between urban density and sustainability (Jenks et al. 1996). Scholars have explored case studies of cities implementing transit-oriented development and smart growth principles, demonstrating their potential to reduce carbon footprints and enhance quality of life (UN-Habitat 2020). Comparative analyses have also shown how cultural and historical contexts influence the adoption and adaptation of planning policies and morphological guidelines across different regions.

6.26 Case Study—Urban Planning Policies and Morphological Guidelines in Patna Municipal Corporation

The primary objective of this survey is to assess public awareness, understanding, and perceptions regarding urban planning policies and morphological guidelines in Patna, as well as to identify the challenges, opportunities, and key factors influencing the city's urban development. A sample of 300 respondents will be surveyed. The sample will be selected using a stratified random sampling method to ensure a diverse representation of the population across different demographic (Table 6.4). The survey is designed as a structured questionnaire containing closed-ended questions. The questions are divided into various sections to capture information on participants' awareness, perceptions, and suggestions regarding urban morphology, policies, challenges, and priorities for urban development.

The survey conducted on urban planning policies and morphological guidelines in Patna presents key insights into the city's urban dynamics. Among the respondents,

Table 6.4 Survey response on urban planning policies and morphological guidelines in Patna municipal corporation

Question	Response options	Number (N)	Percentage (%)
Are you familiar with the urban planning policies in Patna?	Yes	180	60
	No	120	40
Do you think urban morphology affects urban planning in Patna?	Yes	210	70
	No	90	30
How well do you think the urban morphology of Patna aligns with current urban policies?	Well	90	30
	Poorly	150	50
	Neutral	60	20
Do you think current urban policies in Patna adequately address morphological challenges?	Yes	60	20
	No	210	70
	Unsure	30	10
How important is it to incorporate urban morphological guidelines in future policies in Patna?	Very important	180	60
	Somewhat important	90	30
	Not important	30	10
Do you believe integrating morphological guidelines into urban planning would improve the quality of life in Patna?	Yes	240	80
	No	30	10
	Unsure	30	10
Which factor should be prioritized in urban morphology for Patna's development?	Green infrastructure	120	40
	Affordable housing	90	30
	Sustainable transport	60	20
	Resilient urban spaces	30	10
How frequently do you think urban morphological guidelines should be reviewed or updated in Patna?	Annually	120	40
	Every 3–5 years	90	30
	Every 5–10 years	60	20
	Never	30	10

Source Primary survey

60% are familiar with the urban planning policies in Patna, while 40% are not, indicating a reasonable level of awareness but also highlighting a need for increased public education on urban planning. A significant 70% of participants believe that urban morphology directly affects urban planning, reinforcing the importance of physical city structures in policy formulation.

However, when asked how well urban morphology aligns with current urban policies, 50% of respondents indicated that it aligns poorly, with only 30% feeling it aligns

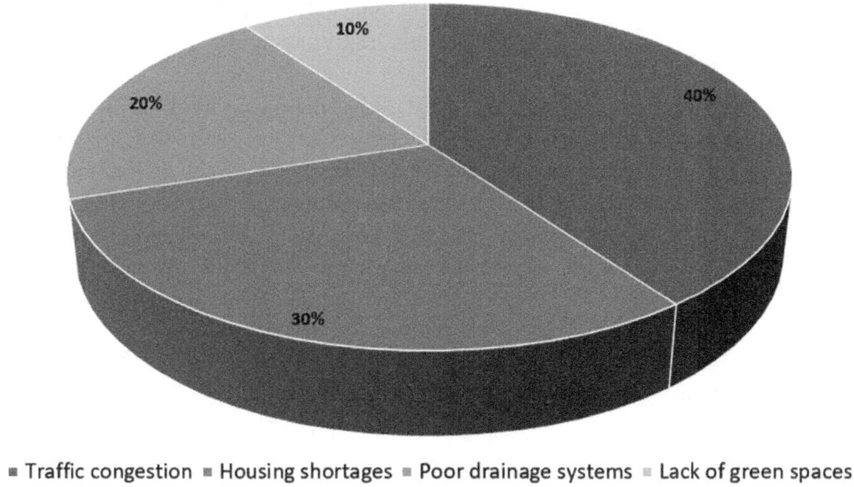

Fig. 6.13 Urban issue in Patna influenced by morphology

well. This suggests that there is a substantial gap between the urban form of Patna and the existing policies, and there is a need for better integration of morphological considerations into the planning process (Fig. 6.13). Key urban issues influenced by morphology include traffic congestion (40%) and housing shortages (30%), with less attention given to poor drainage systems (20%) and lack of green spaces (10%).

In terms of addressing morphological challenges, 70% of participants feel that current urban policies in Patna are inadequate, with only 20% believing they address the challenges effectively. This highlights a clear disconnect between the policies in place and the morphological issues faced by the city. Respondents strongly emphasize the importance of integrating urban morphological guidelines in future policies, with 60% rating it as very important.

The survey also identifies key morphological challenges in Patna, such as overcrowding (40%) and unplanned urban sprawl (30%), suggesting that the city's rapid growth and lack of planning are major issues (Fig. 6.14). A strong majority (80%) believe that incorporating morphological guidelines would improve the quality of life in Patna, highlighting a broad consensus on the potential benefits of such integration.

Respondents prioritize green infrastructure (40%) as a key factor for the city's development, followed by affordable housing (30%) and sustainable transport (20%). Lastly, 40% of participants feel that urban morphological guidelines should be reviewed annually, while 30% suggest a review every 3–5 years, indicating a preference for regular updates to adapt to the evolving needs of the city.

These findings point to a clear need for more effective integration of urban morphology into Patna's urban planning policies, with an emphasis on addressing the city's growth challenges, improving infrastructure, and ensuring that policies are regularly updated to reflect current and future needs.

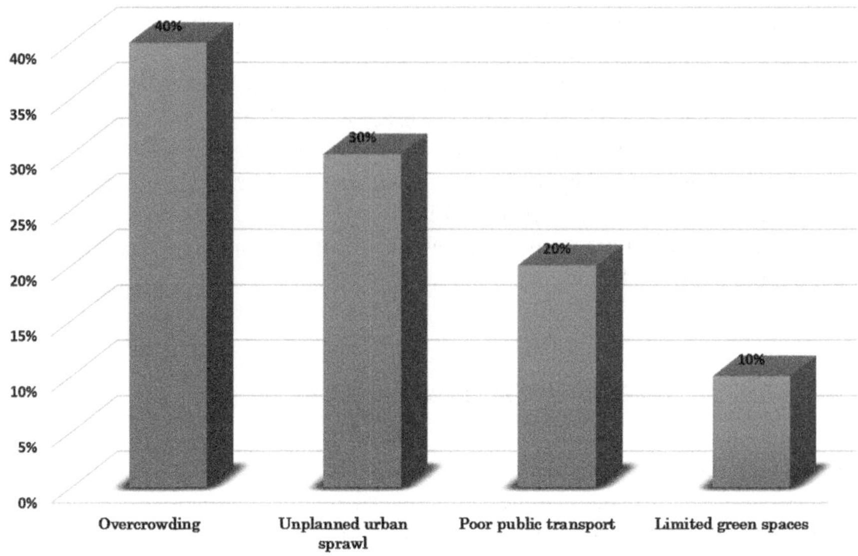

Fig. 6.14 Survey response on key morphological challenges in Patna Municipal Corporation

6.26.1 Role of Governance in Smart City Morphology

Governance plays a critical role in shaping the morphology of smart cities by facilitating the integration of technology, sustainable development principles, and community needs into urban design. Effective governance frameworks ensure that smart city initiatives align with broader urban planning goals, promoting equitable access to resources, efficient infrastructure, and enhanced quality of life. By fostering collaboration among public, private, and community stakeholders, governance mechanisms can drive innovation while addressing challenges such as data privacy, digital equity, and environmental sustainability (Kitchin 2014). Transparent and participatory governance also enables adaptive decision-making, ensuring that smart city morphology evolves in response to technological advancements and societal demands.

Research on smart city governance emphasizes the need for robust policy frameworks and institutional arrangements to manage the complexities of technologically driven urbanization (Caragliu et al. 2011). Scholars have highlighted the importance of public–private partnerships (PPPs) in financing and implementing smart infrastructure projects, as well as the role of regulatory frameworks in ensuring data security and ethical AI applications (Batty et al. 2012). Case studies of cities like Singapore, Barcelona, and Amsterdam demonstrate the impact of participatory governance in shaping inclusive smart city designs (Hollands 2008; Albino et al. 2015). Additionally, literature on urban morphology underscores the significance of data-driven approaches in optimizing spatial planning, highlighting the interplay between governance strategies and urban form (Townsend 2013).

6.26.2 *Stakeholder Participation in Morphology-Based Planning*

Stakeholder participation is a cornerstone of morphology-based urban planning, ensuring that the spatial and structural design of cities reflects the needs and aspirations of diverse communities. By involving stakeholders such as residents, businesses, policymakers, and urban planners, participatory approaches foster inclusivity, transparency, and consensus in decision-making (Innes and Booher 2004). This collaboration enhances the effectiveness of morphology-based planning by integrating local knowledge, addressing socio-cultural dynamics, and aligning development goals with environmental sustainability (Davidoff 1965). Such participation also helps anticipate and mitigate potential conflicts, ensuring that urban forms are functional, equitable, and resilient to future challenges.

Literature on stakeholder participation in urban planning emphasizes its role in fostering democratic governance and community ownership of urban transformation projects (Arnstein 1969). Scholars argue that participatory methods, such as charrettes, public forums, and collaborative mapping, are crucial for bridging the gap between top-down policies and ground-level realities (Healey 1997). Studies highlight the success of participatory models in cities like Curitiba and Freiburg, where active stakeholder engagement has contributed to sustainable urban forms (Albrechts 2004). Research also points to challenges such as power imbalances, lack of resources, and conflicting interests, underscoring the need for structured frameworks to ensure equitable participation (Forester 1999). The intersection of urban morphology and participatory planning is increasingly recognized for its potential to create adaptive, human-centred urban environments (Sandercock 2003).

6.27 Case Study—Stakeholder Participation in Morphology-Based Planning in Patna Municipal Corporation

The survey results indicate a strong belief in the value of stakeholder participation in morphology-based urban planning, but there is significant concern regarding the current state of its implementation in Patna. While most respondents understand its importance, a majority feel that current levels of participation are inadequate and that marginalized groups are often excluded. Additionally, many respondents feel that stakeholder input is underutilized in influencing planning outcomes. The findings suggest a need for more inclusive, effective, and transparent participation processes to improve urban planning and morphology in Patna.

The survey findings provide a comprehensive understanding of stakeholder participation in morphology-based urban planning in Patna. A majority of respondents (60%) are aware of the concept of stakeholder participation, signalling a solid foundational understanding. However, there is a notable gap in the recognition of the importance of this concept, as 80% of respondents believe it is crucial for effective urban planning, highlighting a strong consensus on its significance (Table 6.5).

When it comes to the actual level of stakeholder participation, responses indicate that only 20% view it as high, with 50% rating it as moderate and 30% as low. This suggests a disconnect between the perceived importance of participation and its practical implementation, pointing to the need for greater engagement. The survey also highlights that local residents (40%) should have the most influence in planning decisions, followed by urban planners (30%), with government officials and private sector developers being seen as less influential.

The frequency of public consultations is another key area, with 40% of respondents advocating for annual consultations, while 30% suggest consultations every 3–5 years. The survey also reveals that marginalized groups (60%) are not adequately considered in urban planning, which is a significant concern for inclusivity and equity in decision-making processes.

Regarding the effectiveness of stakeholder participation, 50% of respondents believe it is somewhat effective, and 30% view it as very effective, reinforcing the idea that participation does lead to better urban planning outcomes. Most respondents (80%) agree that stakeholder participation improves urban morphology, further emphasizing its positive impact on shaping the city's development.

The preferred methods for involving stakeholders include public meetings (40%) and online surveys (30%), with workshops and social media engagement being less favoured. However, there is a clear gap in communication, as 60% of respondents feel that urban planners in Patna do not adequately communicate the results of stakeholder participation to the public, suggesting a need for more transparency and feedback loops (Fig. 6.15).

Finally, a majority (70%) of respondents believe that stakeholder participation is underutilized in Patna's urban planning, indicating that there is significant room for improvement in incorporating stakeholders into the planning process. Overall, the findings underscore the importance of enhancing stakeholder engagement, improving communication, and ensuring the inclusion of marginalized groups to create a more effective and inclusive urban planning environment in Patna.

6.28 Challenges and Opportunities

Integrating urban morphology with smart city strategies is fraught with challenges but also offers a fertile ground for innovation. One of the key challenges lies in reconciling the static nature of traditional urban forms with the dynamic and fast-evolving technological requirements of smart city systems. Legacy infrastructure, often designed without considerations for modern technological needs, poses compatibility issues that require extensive retrofitting. Fragmented governance structures and a lack of coordination between urban planners, policymakers, and technologists further exacerbate the difficulty of creating cohesive smart city frameworks. Additionally, the

Table 6.5 Survey response on stakeholder participation in morphology-based planning in Patna municipal corporation

Question	Response options	Number (*N*)	Percentage (%)
Are you aware of the concept of stakeholder participation in urban planning?	Yes	180	60
	No	120	40
Do you think stakeholder participation is important in morphology-based urban planning?	Yes	240	80
	No	30	10
	Unsure	30	10
How would you rate the level of stakeholder participation in current urban planning practices in Patna?	High	60	20
	Moderate	150	50
	Low	90	30
Which stakeholders do you think should have the most influence in morphology-based planning in Patna?	Local residents	120	40
	Urban planners and architects	90	30
	Government officials	60	20
	Private sector developers	30	10
How frequently do you think public consultations should be conducted in Patna for morphology-based urban planning?	Annually	120	40
	Every 3–5 years	90	30
	Only when major changes are proposed	60	20
	Never	30	10
Do you think the participation of marginalized groups (e.g., low-income communities, women) is considered in urban planning in Patna?	Yes	90	30
	No	180	60
	Unsure	30	10
How effective do you think stakeholder participation is in improving urban planning outcomes in Patna?	Very effective	90	30
	Somewhat effective	150	50
	Not effective	60	20
Do you believe that stakeholder participation improves the quality of urban morphology in Patna?	Yes	240	80
	No	30	10
	Unsure	30	10
Do you think urban planners in Patna adequately communicate the results of stakeholder participation to the public?	Yes	90	30
	No	180	60
	Unsure	30	10
How much influence do you believe stakeholder input has on final urban planning decisions in Patna?	High	60	20
	Moderate	150	50
	Low	90	30

(continued)

Table 6.5 (continued)

Question	Response options	Number (*N*)	Percentage (%)
Do you think stakeholder participation in morphology-based planning is currently underutilized in Patna?	Yes	210	70
	No	60	20
	Unsure	30	10

Source Primary survey

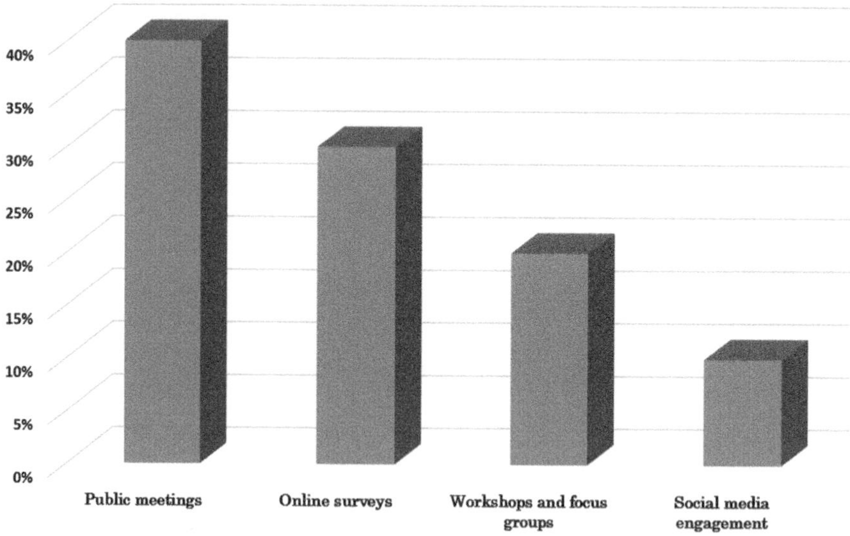

Fig. 6.15 Most effective method for involving stakeholders in morphology-based planning

digital divide and socio-economic disparities in urban areas can lead to unequal access to smart city benefits, undermining inclusivity and equity in urban development.

Despite these hurdles, significant opportunities for innovation in morphological analysis are emerging. Advanced technologies such as Artificial Intelligence (AI), machine learning, Geographic Information Systems (GIS), and the Internet of Things (IoT) are transforming how urban morphology is analysed and understood. These tools enable detailed spatial analysis, predictive modelling, and real-time monitoring of urban systems, allowing for more precise and responsive planning. For instance, AI-driven simulations can optimize land use and infrastructure layouts, while IoT devices provide continuous feedback on urban dynamics, such as traffic flow and energy consumption. Furthermore, the integration of participatory approaches with digital platforms allows stakeholders to engage actively in co-creating urban solutions, ensuring that smart city initiatives align with local needs and cultural contexts. By addressing these challenges with innovative strategies, cities can leverage their unique morphological characteristics to build smarter, more sustainable, and inclusive urban environments.

6.29 Conclusion

Urban morphology is an essential cornerstone in the successful management of smart cities, serving as the foundation upon which technological innovations and urban systems are built. By examining the spatial structure, layout, and form of cities, urban morphology offers critical insights that guide the integration of smart technologies into urban planning and governance. It ensures that the physical environment—such as road networks, building densities, and open spaces—is effectively aligned with the needs of advanced digital infrastructures like IoT networks, GIS platforms, and AI-driven systems. This alignment not only optimizes resource allocation and urban functionality but also facilitates the seamless interaction between urban spaces and the technology that drives smart city operations.

Moreover, urban morphology enables cities to address the challenges of rapid urbanization while safeguarding their cultural and historical identity. By embedding morphological principles into smart city strategies, urban managers can create inclusive and adaptive urban environments that balance growth with sustainability. The role of morphology extends to enhancing livability, promoting equitable access to services, and ensuring resilience against environmental and socio-economic challenges. Importantly, it ensures that the technological advancements of smart cities remain context-sensitive and are designed to complement the unique characteristics of each city.

As cities strive to become smarter and more sustainable, the integration of urban morphology into management frameworks is no longer optional but imperative. It provides a structured approach to harmonize the physical, digital, and social dimensions of urban life, fostering a more cohesive and forward-thinking vision for urban development. By prioritizing urban morphology in smart city management, cities can achieve long-term success in building urban environments that are not only technologically advanced but also socially and environmentally responsible.

6.30 Recommendations

To enhance the role of urban morphology in smart city management, it is essential to adopt an integrated approach that combines spatial planning principles with advanced technological solutions. Policymakers and urban planners should focus on creating frameworks that harmonize physical layouts with digital infrastructure, ensuring that the city's form supports seamless functionality. Leveraging data-driven tools such as GIS, IoT, and AI can provide deeper insights into urban systems, enabling more efficient resource allocation and real-time decision-making. Stakeholder participation should be prioritized to ensure that smart city initiatives reflect the diverse needs and aspirations of local communities. Additionally, standardizing data collection and sharing practices across various systems can foster better coordination and interoperability. Future directions should emphasize equity and sustainability by using urban morphology to design inclusive spaces and environmentally resilient cities. Embracing compact urban forms, green corridors, and mixed-use developments can promote efficiency and sustainability while preserving cultural and historical identities. As cities grow and evolve, continuous innovation in morphological analysis and its application in smart city strategies will be vital for creating adaptive, livable, and forward-thinking urban environments.

References

Akbari H, Pomerantz M, Taha H (2001) Cool surfaces and shade trees to reduce energy use and improve air quality in urban areas. Sol Energy 70(3):295–310

Albino V, Berardi U, Dangelico RM (2015) Smart cities: definitions, dimensions, and performance. J Urban Technol 22(1):3–21

Albrechts L (2004) Strategic (spatial) planning reexamined. Environ Plann B Plann des 31(5):743–758

Alexander C, Ishikawa S, Silverstein M (1977) A pattern language: towns, buildings, construction. Oxford University Press

Angel S et al (2016) Atlas of urban expansion

Anselin L (1995) Local indicators of spatial association—LISA. Geogr Anal 27(2):93–115. https://doi.org/10.1111/j.1538-4632.1995.tb00338.x

Arnstein SR (1969) A ladder of citizen participation. J Am Plann Assoc 35(4):216–224

Arribas-Bel D (2014) Accidental, open and everywhere: emerging data sources for the understanding of cities. Appl Geogr 49:45–53. https://doi.org/10.1016/j.apgeog.2013.09.012

Batty M (2005) Cities and complexity: understanding cities with cellular automata, agent-based models, and fractals. MIT Press

Batty M (2013) The new science of cities. MIT Press

Batty M, Longley P (1994) Fractal cities: a geometry of form and function. Academic Press

Batty M, Axhausen KW, Giannotti F et al (2012) Smart cities of the future. Eur Phys J Special Topics 214(1):481–518

Biljecki F, Ledoux H, Stoter J (2015) An improved LOD specification for 3D building models. Comput Environ Urban Syst 59:25–37. https://doi.org/10.1016/j.compenvurbsys.2016.04.005

Bowler DE, Buyung-Ali L, Knight TM, Pullin AS (2010) Urban greening to cool towns and cities: a systematic review of the empirical evidence. Landsc Urban Plan 97(3):147–155

Caragliu A, Del Bo C, Nijkamp P (2011) Smart Cities in Europe. J Urban Technol 18(2):65–82

Chen J, Li X, Huang B (2020) Machine learning in urban spatial analysis: progress and challenges. Comput Environ Urban Syst 82:101–116

Chelleri L, Waters JJ, Olazabal M, Minucci G (2015) Resilience trade-offs: addressing multiple scales and temporal aspects in urban resilience planning. Environ Urbanization 27(1):181–198

Conzen MRG (1960) Alnwick, Northumberland: a study in town-plan analysis. Institute of British Geographers

Davidoff P (1965) Advocacy and pluralism in planning. J Am Inst Plann 31(4):331–338

Dovey K, Pafka E (2016) The urban morphology of informal settlements. Built Environment 42(2):190–205

Du J, Shi Y, Chen X (2020) Digital twin and urban planning: a systematic review. J Urban Technol 27(3):77–95

Dutta S, Jana N, Ray P (2020) Air quality monitoring using IoT-enabled sensor networks. Environ Monit Assess 192(10):1–12. https://doi.org/10.1007/s10661-020-08569-6

Feng Y, Wang J, Zhang T (2024) The impact of smart city policies on city resilience: an evaluation of 282 Chinese cities. Sustain 16(19):8669. https://doi.org/10.3390/su16198669

Forester J (1999) The deliberative practitioner: encouraging participatory planning processes. MIT Press

Gökçe K (2020) Green infrastructure and urban resilience: a morphological perspective. Urban Studies J 57(6):1204–1221

Graham S, Marvin S (2001) Splintering urbanism: networked infrastructures, technological mobilities, and the urban condition. Routledge

Gungor VC, Sahin D, Kocak T, Ergut S (2013) Smart grid technologies: communication technologies and standards. IEEE Trans Industr Inf 7(4):529–539. https://doi.org/10.1109/TII.2011.2166794

Healey P (1997) Collaborative planning: shaping places in fragmented societies. Macmillan

Herold M, Goldstein NC, Clarke KC (2003) The spatiotemporal form of urban growth: measurement, analysis, and modeling. Remote Sens Environ 86(3):286–302. https://doi.org/10.1016/S0034-4257(03)00075-0

Hollands RG (2008) Will the real smart city please stand up? City 12(3):303–320

Innes JE, Booher DE (2004) Reframing public participation: strategies for the 21st century. Plan Theory Pract 5(4):419–436

Jenks M, Jones C (2010) Dimensions of the sustainable city. Springer

Jenks M, Burton E, Williams K (eds) (1996) The compact city: a sustainable urban form? Routledge

Khajenasiri I, Estebsari A, Verhelst M, Gielen G (2017) A review on Internet of Things solutions for energy control in smart buildings. Energy Procedia 111:770–779. https://doi.org/10.1016/j.egypro.2017.03.239

Kitchin R (2014) The real-time city? Big Data and smart urbanism. GeoJournal 79(1):1–14

Liu Y, Li X, Chen Y (2019) Urban growth simulation using cellular automata and machine learning models. Landsc Urban Plan 187:40–50. https://doi.org/10.1016/j.landurbplan.2019.02.007

Lynch K (1960) The image of the city

Maxwell AE, Warner TA, Fang F (2018) Implementation of machine-learning classification in remote sensing: an applied review. Int J Remote Sens 39(9):2784–2817. https://doi.org/10.1080/01431161.2018.1433343

Meerow S, Newell JP, Stults M (2016) Defining urban resilience: a review. Landsc Urban Plan 147:38–49

Mittelstadt BD, Allo P, Taddeo M, Wachter S, Floridi L (2016) The ethics of algorithms: mapping the debate. Big Data Soc 3(2):205395171667967. https://doi.org/10.1177/2053951716679679

Moudon AV (1997) Urban morphology as an emerging interdisciplinary field. Urban Morphology 1(1):3–10

Neuman M (2005) The compact city fallacy. J Plan Educ Res 25(1):11–26

Oke TR (1988) The urban energy balance. Prog Phys Geogr 12(4):471–508

Pelling M (2011) Adaptation to climate change: from resilience to transformation. Routledge

Perera C, Zaslavsky A, Christen P, Georgakopoulos D (2014) Context-aware computing for the Internet of Things: a survey. IEEE Commun Surv Tutorials 16(1):414–454. https://doi.org/10.1109/SURV.2013.042313.00197

Ramirez Lopez LJ, Grijalba Castro AI (2021) Sustainability and resilience in smart city planning: a review. Sustain 13(1):181. https://doi.org/10.3390/su13010181

Sandercock L (2003) Cosmopolis II: Mongrel cities in the 21st century. Continuum

Steemers K (2003) Energy and the city: density, buildings, and transport. Energy Build

Sun Y, Gao Y, Li H (2022) Identifying socio-economic patterns in urban areas using spatial clustering and machine learning techniques. Urban Studies 59(8):1567–1582. https://doi.org/10.1177/00420980211032076

Taha H (1997) Urban climates and heat islands: alleviating urban heat through vegetation. Atmos Environ 31(15):2615–2626

Townsend AM (2013) Smart cities: Big Data, civic hackers, and the quest for a New Utopia. W.W. Norton & Company

UN-Habitat (2020) World cities report 2020: the value of sustainable urbanization. United Nations Human Settlements Programme

Voogt JA, Oke TR (2003) Thermal remote sensing of urban climates. Remote Sens Environ 86(3):370–384

Zanella A, Bui N, Castellani A, Vangelista L, Zorzi M (2014) Internet of Things for smart cities. IEEE Internet Things J 1(1):22–32. https://doi.org/10.1109/JIOT.2014.2306328

Zevenbergen C, van Herk S, Rijke J (2018) Financing urban flood resilience through public-private partnerships: opportunities and challenges. Water Security 4:1–7

Zhang M, Kukadia N (2005) Metrics of urban form and the modifiable areal unit problem. Transp Res Rec 1902(1):71–79. https://doi.org/10.3141/1902-09

Zhou C, Li X (2025) Impact of smart cities on urban resilience: the roles of land green utilization efficiency and industrial structure transformation. Land 14(7):1373. https://doi.org/10.3390/land14071373

Chapter 7
Future Smart Cities and Sustainable Development

7.1 Introduction

Making the most use of data, technology, and available resources is how smart cities improve urban development, administration, and execution of services, as well as public involvement and accountability. The goal of developing cutting-edge technology for smart cities, such as digital twins, artificial intelligence (AI), big data, and the Internet of Things (IoT), is to provide a civic platform for innovative urban designs that combine sustainability and agile wealth. In other words, although new technologies are crucial facilitators, the foundation of successful smart cities is promoting integration and congruence in urban design and governance.

The production and use of resources needed for residential, industrial, transit, commercial, or recreational operations produces cognizance of sustainability and sustainable urban development principles (Yigitcanlar and Kamruzzaman 2015). The goal of sustainable city planning in smart cities is to raise environmental consciousness in the usage of natural resources (Komninos 2016) (Table 7.1).

7.2 Issues and Challenges in Smart Cities—An Advanced Analysis

Due to the high level of greenhouse gas emissions generated by urban activity, they constitute a "major contributor to climate change." Over 95% of the 234 cities most affected by climate change and classified as "extreme risk" in the Climate Change Vulnerability Index (CCVI) are in Africa and Asia, fuelling fears that the world's poorest countries with the highest rates of urbanization will bear the brunt of climate change's costs. The magnitude of the danger might jeopardize financial flows that have poured into these countries to take advantage of growing economies, growing consumers, and low labour costs.

G. S. Bhunia and P. K. Shit, *Urban Morphology and Sustainable Smart Cities*,
https://doi.org/10.1007/978-3-031-97473-1_7

Table 7.1 Models, frameworks, and tools for sustainable smart city (Trindade et al. 2017)

Model	Key features	Goal	References
Model for computing the smart city indices	Fuzzy logic	To develop sustainable urban model in policymaking process	Lazaroiu and Roscia (2012)
Korea city model, u-eco-city	ICT and eco-technology	To offer residents, employees, and guests a high standard of living and an environment with little to no detrimental effects on the environment	Yigitcanlar, and Lee (2014)
Analytical framework	ICT and energy efficiency	How ICT assets could help cities minimise their energy consumption	Kramers et al. (2014)
Conceptual framework	Conceptual dimension and subdimension	To generate taxonomies of smart city practices	Lee et al. (2014)
Conceptual framework	Diffusion innovation theory	Creating a theoretical framework with a grounded approach for integrating innovation policy	Cohen and Amorós (2014)
Elite cities tool	Low carbon eco-city; indicator-based analysis	Evaluates the status of 33 key indicators chosen to reflect the most important concerns	Zhou et al. (2015)
Eco-city model	Centralized hub, cluster analysis	To establish connections between grid networks, economic relations, and transportation	Hu et al. (2015)
Holistic system dynamics methodological framework	Decision-making and policy implementation	In order to support managers, municipal leaders, and decision-makers in creating and implementing efficient policies for tracking and evaluating eco-city sustainability	Tsolakis and Anthopoulos (2015)
Intelligenter method	Sophisticated architecture, complex mater plan	Innovative idea of collaborations discovery in urban systems	Marsal-Llacuna and Segal (2016)
3Es framework	Efficiency, economy, and effectiveness	Systematic integration of several components of environmental sustainability is necessary to provide a framework for evaluating an eco-city's achievement in sustainability	Hu et al. (2016)
Indicator framework	Low carbon city	To create a framework for assessing an eco-city's sustainability accomplishments, a number of sustainable development elements must be integrated systematically	Tan et al. (2017)

Over the course of the next 30 years, approximately 70 million individuals will move annually to urban areas. Two thirds of the worldwide population will live in cities by 2050. In 2018, CDP received climate and environmental data from over 620 cities. Climate hazards were documented in 530 of these cities, with a combined population of 517 million people. Cities are already suffering the effects of these dangers, with 42% of identified hazards projected to materialize in the near future. Because the majority of these dangers (60%) are rated as having a medium or high chance, communities may expect to suffer the full brunt of climate change in the future years. With statistics showing that by 2050, eight times the number of city inhabitants would be exposed to high temperatures and 800 million more people will be at risk from the consequences of increasing sea levels and storm surges, cities are unable to afford to disregard medium-long-term dangers. Warming above 1.5 °C will result in "exacerbated urban heat islands, amplification of heat waves, extreme weather volatility, floods, droughts, coastal inundation, and an increase in vector-borne diseases like malaria and dengue fever," according to the IPCC's recent Summary for Urban Policymakers on the 1.5 °C report. Despite this, only 11% of risks are identified as long-term.

7.2.1 *Climate Change and Social Risk*

Due to extraordinary urban expansion, which is projected to keep going in the next decades, cities' vital social care and facilities are presently under stress. These social and economic issues already exist, and climate change will only make them worse because of the physical disruptions it is predicted to cause to business and government operations. Cities also report a significant number of present and short-term societal dangers, but longer-term concerns are reported less frequently in all parts of the world. While we anticipate an increase in social hazards in the next decades, cities are only citing "population displacement" as a major threat. Cities can only plan infrastructure and services effectively if they have a good understanding of long-term risks.

'Greater danger to already vulnerable groups' is one of the top social worries connected to our changing environment, which is not unexpected given that the poorest people of society would be the most affected by climate change's uneven distribution of effects. Furthermore, every half-degree additional warming will worsen public health. In comparison to a 1.5 °C temperature rise, IPCC assessments indicate that a 2 °C temperature rise will end up resulting in a major increase in extreme temperatures, drought, and vector-borne illnesses. Some cities are recognizing this, with 17% citing the risk of "greater demand for public services" and 16% citing the risk of "increasing incidence and prevalence of disease." (Figs. 7.1 and 7.2).

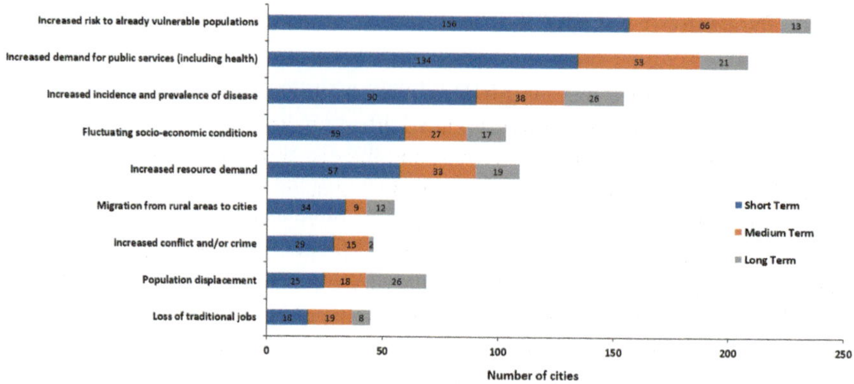

Fig. 7.1 Climate change poses serious social risks (*Source* https://www.cdp.net/en/research/global-reports/cities-at-risk)

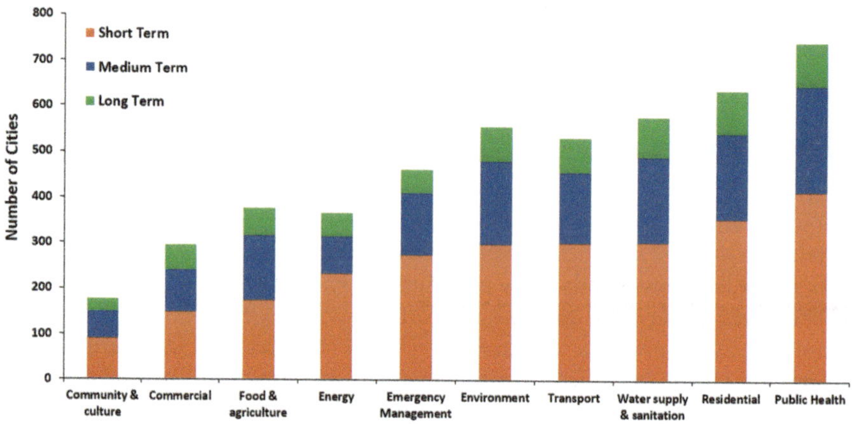

Fig. 7.2 Climate change is threatening vital public services (*Source* https://www.cdp.net/en/research/global-reports/cities-at-risk)

Cities will continue to encounter a variety of issues in the next decades as their populations rise, some of which will jeopardize their ability to handle the consequences they face. Cities, on the other hand, can enhance the welfare of their inhabitants and develop resilience against climate change by engaging in the correct infrastructure and services. Cities that reported to CDP in 2018 cited a wide range of issues that could promote or inhibit their transition to a climate resilient future. Cities are often better at identifying issues than facilitators; 863 obstacles to action were noted compared to 456 enablers. Global cities are seriously threatened by climate change. The world has already warmed by one degree Celsius beyond pre-industrial levels, and even if significant action is taken to minimize additional warming, threats are expected to persist for centuries. Cities must take steps to increase resilience and safeguard inhabitants from the effects of climate change. Cities must first acknowledge their vulnerabilities. Every city should conduct thorough vulnerability assessments that take into account both the threats that are present now and those that are anticipated to arise in the upcoming years and decades.

7.2.2 A Smart City is Green and a Green City is Smart!

Plant life in cities has numerous advantages: it can enhance air quality and drainage, include a source of food, and give a welcome aesthetic contrast to concrete and glass-dominated cityscapes. However, in metropolitan areas, competition for space is severe, with the need for homes and industrial spaces sometimes infringing on the limited green space available. A major focus of the smart city programme is reducing carbon dioxide emissions by using environmentally friendly energy sources, smart electrical supply systems, and improved effectiveness. Nature, in terms of soil, water, ecosystems, and species, is rarely included in smart city planning. In the process of greening the city, digital technology plays a critical role. Drones and RFID tags can facilitate preserving and sustaining plant life cheaper and more efficient, even if it's halfway up a skyscraper, thanks to satellite surveys and machine learning. There is already software that can track weather forecasts and remotely dump rainwater tanks in advance of rainstorms to avoid floods. Such systems might be implemented citywide, lowering flood danger while simultaneously increasing watering rates for roof gardens and other irrigated vegetation ahead of heat waves.

Infrared satellite footage was blended with data from Google Earth by scientists in London. Since plants and trees reflect infrared light uniquely than man-made surfaces like roads and roofs, it was simple to create a map of the city's extant green area. This approach can also be used to discover contenders for urban farms and green roofs, which are popping up all over the world and allow plants to be grown on top of flat structures. For example, Infarm is a modular, app-controlled hydroponic growth system that allows urban farms streamline some formerly laborious procedures. Another German business, Optigreen, has developed an app that permits exact management of the amount of water that drains from a green roof.

Biologists utilize GPS technology to track free-ranging and migratory animals. GPS technology allows organizations to track cars and devices such as cell phones, but it is also used by biologists to track free-ranging and migrating animals. The use of GPS technology in the research of wildlife movement through cities could aid planners in identifying movement impediments and determining the optimal locations for new habitat creation.

Moreover, technology has the potential to assist municipal governments in maintaining their current parks and green spaces. While most smart city programmes have concentrated on transportation infrastructure or the physical environment, "smart parks" are increasingly being discussed. Green space is becoming more widely recognized as an important factor of happiness, and several projects aim to encourage citizens to make better use of whatever greenery they have access to. The idea of making the city more transparent to both wildlife and people could be aided by putting in sensors that track wildlife movement. Several cities have already begun to plan for climate change adaptation utilizing natural treatments. Additionally, the efficiency of urban green varieties and permutations in producing cooling and absorbing rainfall will be significantly improved by extensive and widespread real-time observations of temperature, humidity, evapotranspiration rate, and flows. By integrating this data with maps of suffering and lack, cities will be able to evaluate greening activities more intelligently (Figs. 7.3, 7.4 and 7.5).

- ***Case Study 1***: Smart city activities in Singapore

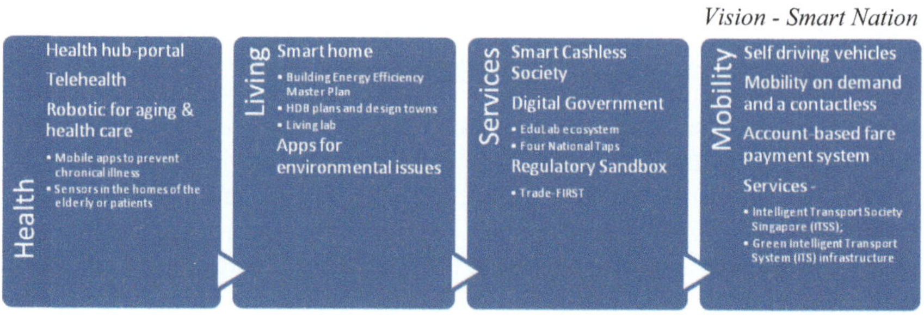

Fig. 7.3 Smart city activities in Singapore

Fig. 7.4 Smart city activities in Dubai

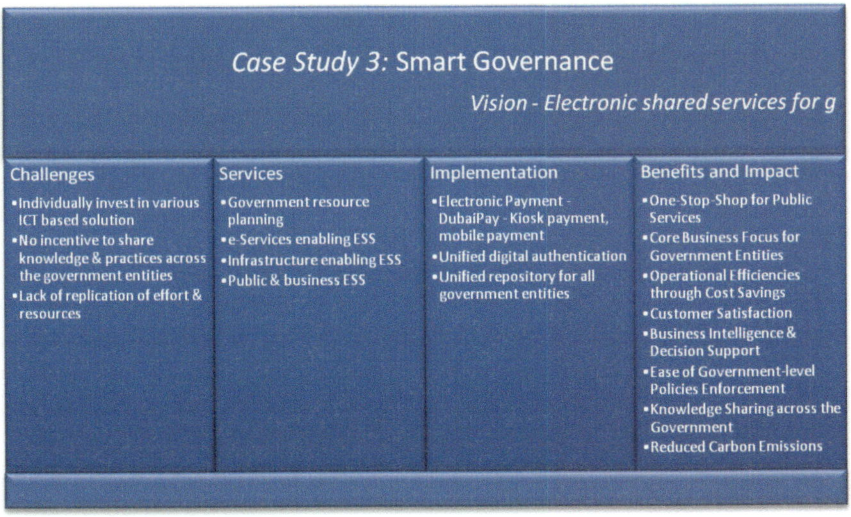

Fig. 7.5 Smart city and smart governance

7.2.3 Waste Management with Smart Solutions and Technology

As a result of population growth and fast urbanization, garbage creation has skyrocketed, rendering traditional waste collection systems ineffective and expensive. By 2050, garbage production worldwide is expected to reach 4 billion tonnes, more than tripling from 2016. The rapid rise can be attributed to the recent rise in urban populations and consumer culture, neither of which is predicted to slow down any time soon. To reduce their impact on the environment and garbage collection products and services, communities worldwide are resorting to intelligent waste management technologies and solutions. The goal of smart waste management is to address the solid waste management issues mentioned above by utilizing applications for mobile devices, intelligent monitoring systems, and sensors. The first intelligent waste management device to improve the effectiveness of the rubbish collection operation is the sensor. In addition to providing real-time updates and alerting waste management providers when the receptacles are full or almost full, sensors can track the amount of fill in the receptacles.

When it comes to solid waste management concerns, municipalities may profit from the data intelligence and real-time insights offered by IoT devices. In this instance, specific container fill tendencies may be found using historical data and subsequently tracked over time. Mobile applications, in addition to hardware solutions, are utilized to tackle issues in the traditional waste management system, including maintaining track of drivers while they are on the road. Fleet management software makes it possible to have real-time visibility, operational efficiency, driver behaviour, real-time warning, and much more. Polish company Bin-e developed a smart garbage can that uses AI-based object recognition to automatically divide recyclables into several sections, cutting down on incorrect recycling sorting. After separating the garbage, the machine compresses it while monitoring how filled each bin is. By removing human error from the first sorting process, smart trash bins enable recycling facilities to deal with materials more quickly and effectively. This can increase worker productivity and reduce waste management expenses by up to 80% (Fig. 7.6).

A reduced staff during the COVID-19 epidemic has left several institutions unable to keep up with demand. Fortunately, recycling robots with artificial intelligence (AI) capabilities can help pick up the slack. With their ability to accurately identify and classify recyclables, these robots increase productivity and reduce the need for human labour. In the long run, this helps recycling facilities avoid paying landfill fees while also saving money. Technology for garbage truck weighing can reduce the number of pickup trips and assist predict fill levels. They accomplish this by caching and weighing garbage containers, which they subsequently utilize to predict fill levels over time.. Cities might reduce total collections costs by using this technology to more accurately determine how frequently they should send trucks out. There is an increasing need for waste management systems in cities to keep up with the growing volume of waste generated. In order to solve this problem, a number of communities are installing pneumatic garbage disposal units that are connected to a system of underground pipelines. Pipes are used to move trash to a waste collection centre, where it is either pulled away or treated. This approach reduces energy expenses, increases efficiency, and substitutes routine garbage pickup. To increase collection efficiency and save trips to and from the disposal, Ecube Labs developed a solar-powered garbage compactor that can hold up to five times more waste than standard trash cans. These devices record and transmit data on fill and pickup times, as well as compress waste as it accumulates to increase bin capacity and streamline collection procedures. Carelessly discarding electronic waste may be harmful to both people and the environment. Thankfully, a large number of companies and groups have started e-waste recycling initiatives that will take in your outdated electronics and even pay you for them. A firm focused on smart recycling called ecoATM created a range of e-waste recycling kiosks that allow you to instantly exchange your electronics for cash (Table 7.2).

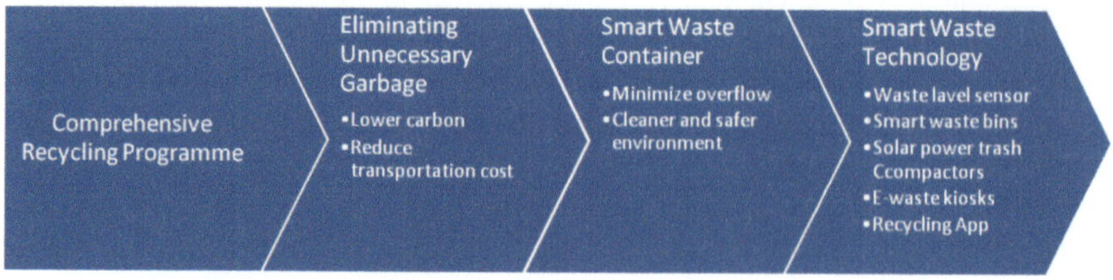

Fig. 7.6 Schematic diagram of smart waste management

Table 7.2 Case studies of smart waste management

Smart waste management technology	Geography	Benefits
AI-based robots	San Francisco, California	• Diverts around 80% of its garbage from landfills • Quickly and accurately sort recyclables
Truck-free waste management system—underground pneumatic waste pipes	Songdo, South Korea	• Automatically sorted garbage is recycled, buried, or incinerated for energy • Cut down on carbon emissions
Weighing mechanisms: • Fill-level sensors in public trash bins • Floating dumpsters in canals	Amsterdam, the Netherlands	• In historical sites, abolish necessity curbside garbage pickup and replace it with a more energy-efficient option • To develop waste collection plans that are more efficient and cost-effective

7.2.4 *Striving for Carbon–Neutral Cities*

The eco-city places a greater emphasis on the environmental elements of sustainability, such as the physical environment and ecosystems, than on the economic and social aspects (Holmstedt et al. 2017; Rapoport and Vernay 2011). While the surrounding habitat has always been a major concern in urban planning (Wong and Yuen 2011), ecocities put it to the forefront of the strategic planning. As local governments amass location-specific information, residential information, citizen knowledge, and historical data electronically, the amount of data being collected in cities is continuously expanding (Wammes 2015). The challenges that urban policymakers face is mostly due to the information and communication management of data collected on an hourly or even second basis, rather than the technology itself (Flowers 2013). According to studies, technological advancements have resulted in the collecting of numerous environmental data. Many of these statistics are gathered for specific programmes or efforts at the national, subnational, and local levels. Incompatibilities in formats, standards, access, and organizational structures make integrating the data difficult. According to research, technological advancements have resulted in the collecting of numerous environmental data. Many of these statistics are gathered for specific national, subnational, or municipal projects or initiatives. Incompatibilities in formats, standards, access, and organizational structures are preventing the data from being merged. Finally, without effective data integration and analysis, the obtained data cannot be turned into decision-making knowledge (Copeland 2014). This necessitates not just a data ecosystem capable of capturing, storing, processing, managing, transmitting, and sharing data (Gant 2015), but also, from a government standpoint, ways to comprehend, transfer, and use the knowledge generated by big data. According to Bibri (2020), there are several eco–city models

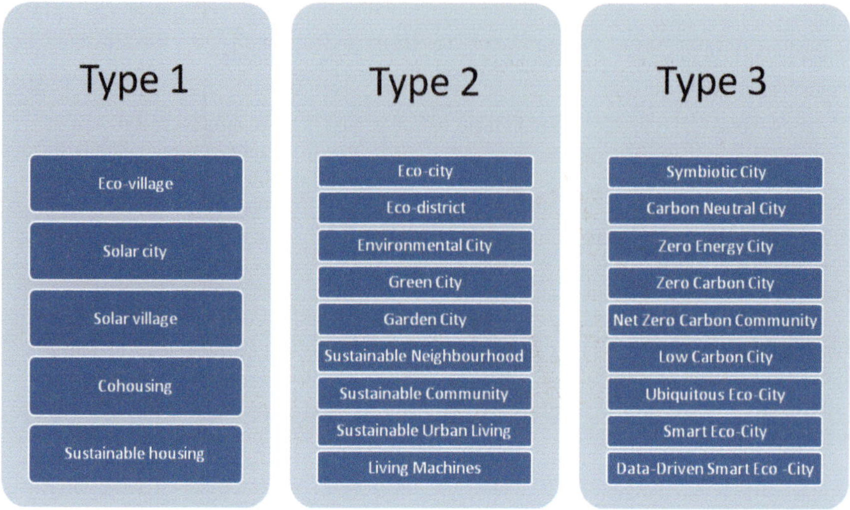

Fig. 7.7 Three types of eco-city models

that can be classified into three categories: Type 1 focuses on passive solar design, type 2 on passive solar design and greening, and type 3 on green energy technologies and/or smart energy and environmental technology (Fig. 7.7).

To have the highest air quality, a city must drastically help Mitigate climate change from its energy and transportation sectors, as well as become carbon–neutral. One of Barcelona's main objectives, as stated in the Municipal Action Programme, is to create a new development model for an emissions-free, hyperconnected, and healthful city "where the environment, urban planning, and ICT infrastructures are fully integrated" and distinguished by "productive neighbourhoods at a human pace" (Mora and Bolici 2016; Bibri 2020). The city of Stockholm wants to leverage digitization and new technology to find things simpler for citizens and businesses to minimize their energy use and carbon footprint, and so become more environmentally conscious (Stockholm City 2017). Stockholm's eco-development SRS is presently testing smart power grids (and the combination of district heating with low-energy dwellings) as specified in the Master Plan for Stockholm (Stockholm City 2018). Brandt and Nordström (2011) show how the SRS-M Information Management System may be used to combine 150 criteria spanning electricity, district heating, transportation, water, waste, and a variety of other environmental and social aspects (SRS-M IMS). The researchers show that the smart grid as the anchor of smart cities only completely permits less than 15% of the 150 indicators specified in the SRS-M, relying on a functional gap analysis. From innovating energy–efficient engineering solutions in construction and infrastructure to the implementation of intelligent electricity networks that enable local production and distribution of electricity, it provides tremendous prospects for climate–adapted and future–oriented prosperity (Stockholm City 2009).

The CPH 2025 Climate Plan was approved by Copenhagen City Council in 2012. A major element of this strategy is the implementation of smart city projects, which aim to make Copenhagen carbon neutral by 2025. Copenhagen joined the IBM Smarter Cities Challenge to use data for policymaking in tandem with its decision to use data as a means to minimize energy use (IBM 2013). Following that, the city teamed up with IBM to collect energy data in order to reduce CO_2 emissions. IBM is currently gathering and directing important data into an "open data hub." "The goal is that the hub will connect data suppliers and users, as well as entrepreneurs and programmers, so that they can utilize the information to enable the city save energy" (Online Post 2013). Additionally, the city and Hitachi Consulting are working together to build a "City Data Exchange" Hub where information may be stored and accessible by the general public or bought by other businesses. "By combining data from numerous sources, the platform is expected to enable advanced analytics to support city activities such as green infrastructure development, traffic control, and energy usage" (Hitachi Consulting 2015).

7.2.5 *Green Infrastructure and Smart City*

Cities account for 2% of the total geographical area on the planet and emit 70% of total CO_2 emissions (UN Habitat 2011). CO_2 emissions are absorbed by forests to the tune of 40% (Gray 2011). Extreme weather brought on by climate change poses a significant infrastructure issue for cities around the world, and much research has been done on the subject. Consequently, increasing the number of plants and trees in a city is essential. The nexus of smart, green, and grey technologies is evolving. These technologies are able to adjust in response to stakeholder reactions to climate change and increasing environmental consciousness because of the development of innovative technical systems. The term "green infrastructure" alludes to infrastructure that is eco-friendly and contributes to city growth that is long-term. Natural habitats and wetland characteristics, for example, can help to lessen the danger of coastal and riparian flooding (UK Green Building Council 2015). Heavy rain will interrupt transportation and increase sewage floods. The proportion of assets vulnerable to coastal flooding will expand when sea levels rise as expected (Committee on Climate Change 2017). To struggle the noteworthy effects of climate change on infrastructure, the government reviews the risks every five years. Only by adopting energy-efficient solutions, we can achieve sustainable development. Green infrastructure includes things like botanical gardens and parks, natural as well as semi-natural metropolitan green areas, and green passageways. Green infrastructure is designed to provide not only commercial but also environmental benefits. Green landscapes not only beautify, but also naturalize and complement metropolitan areas. Green cover has numerous advantages, extending from temperature decrease to wind flow modification. Construction of energy-efficient buildings, efficient use of power and water, and clean air must be prioritized by the government, urban planners, and other stakeholders. Nonetheless, given the different urban development authorities and municipalities that operate in Indian cities, securing green infrastructure investment might be difficult. There must be public awareness campaigns that convey the advantages of green infrastructure to all parties concerned. Green infrastructure investments require new norms and conventions, just as traditional infrastructure investments require them.

The term "grey infrastructure" describes constructed assets that provide services, such as transportation, rehabilitation centres, stormwater and sewage systems, or storage basins, and are essential to community operations. The word "grey" relates to the fact that concrete is frequently used in such structures (NGICP and IGICP 2018). The purpose of grey infrastructure is to supply critical urban operations. However, they obstruct natural processes such as surface water flow and filtration, animal migration, the food chain, and plant succession (Benedict and McMahon 2002). Built infrastructure, sometimes known as grey infrastructure, is intended to serve a single specified utility, such as drainage or transportation (Liquete et al. 2015). An interconnected network of green spaces is managed and controlled to provide a wide range of environmental, social, and financial benefits, including urban parks, private gardens, green walls and roofs, urban wetlands, and community and tenement plots (Foster et al. 2016). The "smart city" agenda aims to leverage technology, big data, and data platforms to aid decision-making and promote effective operational networks in cities (Woetzel et al. 2018). Platforms such as smart city systems can be applied across multiple sectors, including intelligent pricing for parking, transit, energy, and water; as well as the management of energy grids, drainage networks, communication systems, and transportation. Smart infrastructure services are the name given to these industries (Fig. 7.8).

Smart cities are facilitating the next wave of infrastructure expenditures around the world (Woetzel et al. 2018), with more than $20 trillion in incremental economic gain expected by 2026 (Woetzel et al. 2018; Smart Cities World 2018). There have been claims that instead of fostering inclusive communities, smart cities—especially

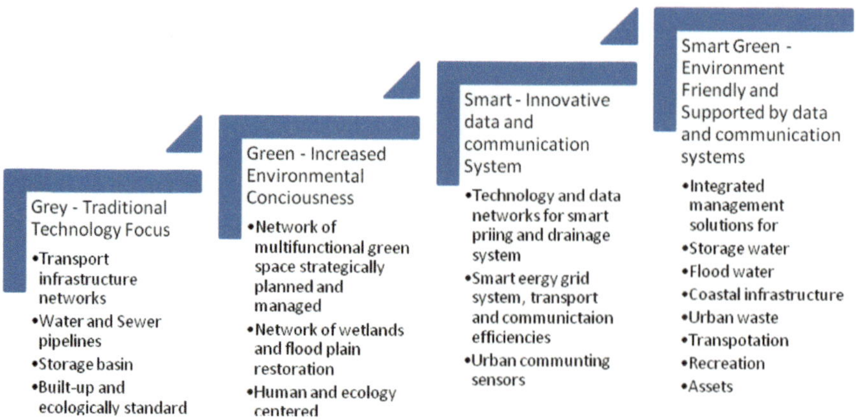

Fig. 7.8 Concept of green infrastructure

Table 7.3 Integrated smart and green nature-based solutions for future cities

Geographic region	Solution	References
Masdar City, Abu Dhabi	Zero waste, zero carbon, car-free, fed by recycled water	Liu (2018)
Songdo, South Korea	Networked computers controlling all aspects of life; lonely and transparent city	Albert (2019)
Eko-Atlantic, Nigeria	Waterfront reclamation	Albert (2019)

those that have just been established—have isolated individuals. Many cities are known for their great regard for technology and for lacking many of the elements that improve the standard of living in residential communities. In order to combat climate change and ensure long-term sustainability, cities need to be future-proofed. Infrastructure that is adaptable and responsive will be required for this. In order to enable the making of intelligent judgements, this need ensures that systems, such as networks, sensors, and communication channels, are closely watched over and that data is effectively gathered and distributed. The findings point to a shift away from silo infrastructure and towards interconnected smart and green infrastructure systems that improve the environment, health, and economy. In order to bring these interdependent and linked systems to market, stakeholders need to be convinced that integrated smart green solutions can be more effective than isolated ones. Thanks to data-driven innovations like Big Data and IoT, interconnected infrastructure systems will be able to create future cities that will support smart government, smart living and neighbourhoods, intelligent transportation, and smart economies and ecosystems (Table 7.3).

7.2.6 Improving Eco-efficiency

Nevertheless, an effective energy network must complement our connection infrastructure in order to enable sustainable smart cities in the future. In addition to saving money, smart towns need to become more energy efficient in order to help the nation reach its net-zero carbon emissions goals. As certain groups are suspicious of further technology integration in urban design, smart cities need to immediately prove their value by reducing energy expenses and resulting in real advancements in daily living. In addition to fostering community engagement, this will position Britain as a global leader in green technology. As energy enters the digital era, protecting data anonymity and security will be critical to smart cities' long-term success. This commitment must be incorporated in government policy and fulfil the highest security standards: safe data will result in a more sustainable system that can be trusted to further environmental conservation projects. One smart energy network will be needed to enable this growth if smart cities are to be widely adopted and lead to significant savings in the energy usage of densely populated metropolitan settings. The Internet of Devices

Table 7.4 Green IoT for eco-friendly and sustainable smart cities (modified after Almalki et al. 2021)

Smart technology	Reduce energy	Reduce CO_2	Reduce cost	Climate change
Data centre	√	√	√	√
Wireless sensors	√	√	√	√
Cloud computing	√	√	√	√
Communication technologies	√	√	√	√
GIS	√	√	√	–
ICT-based solid waste management	√	√	√	√

refers to the connectivity between things in a smart city (IoT). IoT encompasses everything in smart cities that can be connected at any time, from anywhere, and through any medium (Minerva et al. 2015). IoT technology advancements continue, making IoT components smarter through an adaptive communication network, processing, analysis, and storage (Perera et al. 2014). Cameras, sensors, Radio Frequency Identification (RFID), actuators, drones, mobile phones, and other IoT devices are examples. All of these have the potential to interact and collaborate in order to achieve common goals. As a result, the integration of IoT and practical approaches to minimize the power consumption of big data processing and transmission can enhance the quality of care in smart cities while also contributing to getting the world greener, more ecological, and a safer place to live (Syed et al. 2020; Alsamhi et al. 2019). Additionally, green IoT solutions include lowering CO_2 emissions and lowering IoT energy consumption in order to complete the smart world with the sustainability of intelligent everything. Green IoT encompasses both the design and implementation of environmentally friendly features. The development of computer equipment, energy efficiency, communication protocols, and networking topologies are examples of green IoT design elements (Table 7.4).

7.2.7 Increasing Sense of Place

The term "sense of place" (SoP) relates to the individual rather than the location (Vanclay 2008). Lowery and Morse (2013) define SoP as a multifaceted construct shaped by the emotions, ideas, and actions that people identify with a location (Jorgensen and Stedman 2001). Measuring SoP is a difficult task, particularly when it must be done spatially. In any event, because their affective ties are towards a geographical location, SoP and other place-related notions such as place attachment, place reliance, and place identity are suited to be spatially quantified. Historically, a sense of location was an unintentional emotional state. It was built through long procedures that could not be replicated in the purposeful effort of building a modern metropolis (Kourtit 2021). The citizen-centric smart city strategy is built on interpersonal interaction, which are largely dependent on our ability to grasp spatial (space-based geography that is relevant to human ideologies, social ideals, and human–space interactions) urban dynamics (Roche 2016). Although communities, governments, and corporations may all benefit from the notion of urban kinetics, the study's primary emphasis is on individual citizenship. The concept of urban intelligence is directly linked to the effective implementation of those interpersonal communication. This research focuses on individual citizenship, even if the notion of urban dynamics may be extended to enterprises, governments, and communities. The concept of urban intelligence is directly linked to the effective implementation of those interpersonal communication. Nowadays, we use the sensing part of urban engineering (i.e. sensors, location-based and context aware services) to grasp a city's urban intelligence, but the dilemma is to go one step further and perceive individual specialities (individual specialities in these studies are defined by Lussault (2007) as collective and individual practice in relation to their geographic locations and to one another that portray their spatial activities. As a result, there are two distinct perspectives on the environment of smart cities: one that is founded on urban engineering and its location-based technology paradigm as (dynamic) layers throughout the city, and its social parallel, which is a representation of the city built on the dynamism of urban intelligence as a network of locations.

Nowadays, we were using the detecting part of urban engineering (i.e. sensors, location-based and enable new services) to grasp a city's urban intelligence, but the dilemma is to go one step further and decipher individual spatialities (individual spatialities in this research are defined by Lussault (2007) as individual or collective practices in relation to their geographical location and one another that represent their geometric behaviour). The interpretation of data with qualitative meanings and feelings has been a difficulty for Geographic Information Systems. The latter concerns are partly attributable to place's spatial ambiguity (Jones et al. 2008) and dynamism (Roche 2016) in comparison to the Euclidean representation of space. Indeed, Relph (1976) suggested that location is not a requisite of place, but Cresswell (2004) contended that place is never complete; rather, it is always evolving. The intrinsic fluidity and border vagueness of complex notions such as location, SoP, and SC, as well as the key consequences of mapping through GISc techniques, appear to be crucial reasons why these ideas are not operationalized. As a process of social relationships, Massey (1994) has already outlined the inherent issues of conceiving limits for the place idea. Moreover, using geographic primitives (e.g. discrete points and/or polygons) to explain complex and multifarious notions (e.g. location, SoP, SC) can result in a number of challenges and information loss. Brown et al. (2015) were the first to seek to quantify and map the concept of place attachment. They employed an Internet-based public participatory Geographic Information System (PPGIS) to obtain all of the essential data, based on home range conceptualization (Powell and Mitchell 2012). Jorgensen and Stedman (2011) used structural equation approaches to evaluate the spatial component of sense of place by combining geographic and physical aspects of places with attitude and behavioural factors. Jenkins et al. (2016)

recently combined Twitter data with VGI from Wikipedia to spatialize a collaborative SoP using social network analysis (SNA).

7.2.8 Sustainable Transport

According to Dia (2018), traffic congestion costs 1–3% of a country's GDP on average, while comprehensive road death rates reach 1.2 million and cost USD $100 billion each year. To sum up, contemporary cities have gone above and beyond to construct intelligent transport networks that enhance both the mobility of their citizens and the flow of urban traffic. Several urban centres have just begun to shift into the digital age through the intentional integration of information and communication technologies into transportation and administration in general. According to Transport Canada (Dia 2018), contemporary transportation systems necessitate high consumption rates of both renewable and non-renewable resources. Zhou (2012) also emphasized the need for smart transportation in the light of the impending threat posed by global warming, which is mostly brought on by emissions from motor vehicles. As a result, smart transport enhances basic in-city travel while enhancing the standard and functionality of transit inside any given modern metropolis. Furthermore, Olaverri (2018) found that smartphone devices might be used as a communication platform to notify pedestrians and vulnerable road users on public transportation systems about their itineraries, hence reducing dangers.

Sustainable transportation is one of the key engines driving development, and it has been for a long time. Sustainable transportation is incorporated into the United Nations' Sustainable Development Goals (SDGs), most notably SDG 11, which asks for inclusive, safe, resilient, and green cities and human settlements (Bamwesigye and Hlavackova 2019). Due to worldwide lockdowns implemented in reaction to the COVID-19 epidemic, mobility decreased in 2020 in a way never seen before. The IEA reports that as of the end of March 2020, the global average of road transport activity had decreased by around 50%, while road transport reduced by 50 to 75% in areas under lockdown. However, global carbon dioxide emissions have increased, with December 2020 emissions 2% greater compared to December 2019 concentrations. AI, IoT, and cloud computing are examples of new innovations that can add intelligence to the transportation business. Cities may use intelligent transportation systems (ITS) to help reduce traffic congestion and pollution, reduce fuel and electricity consumption, and improve travel times by emphasizing the value of these new technologies. For public transportation, availability might be adjusted based on trip demand, and passengers could be given information to help them plan their routes. This will necessitate the incorporation of these capabilities into a citywide infrastructure capable of monitoring and managing traffic conditions as well as developing forecasting models to anticipate potential congestion. Communications will be required between automobiles and infrastructure, which may raise privacy and security problems.

A number of cities, including Helsinki, Santiago de Chile, and Kolkata, have embraced electric buses in order to improve pollution levels. A record number of electric trucks were shipped worldwide in 2019, and recent research on dynamic charging methods has broadened the range of long-haul transportation. New battery solutions are also hastening the electrification of transportation. In early January 2021, for example, an Israeli business reported that it had developed batteries that could be fully charged in five minutes. Complementary batteries are being developed by other companies throughout the world, and it is predicted that this technology will be accessible to the general public within the next five years. The next stage in public mobility is to integrate the transportation system into the electrical grid. The energy grid is in the process of incorporating smart technology and renewable energies into the network. This will allow electric vehicles to get electricity from clean energy sources, decreasing pollution even further. According to the IEA energy can be stored in electric car batteries and then provided to the grid at appropriate times via vehicle-to-grid systems (V2G).The IEC Systems Committee for Smart Energy oversees the efforts of many technical committees (TCs) attempting to establish standards related to grid digitalization, automation, and modernization, comprising grid edge devices and systems.

The majority of modern cities are eventually intended to be driven through. As a result, in this scenario, planning is mostly for long distances, making bicycle transportation difficult in such places. Therefore, any community that chooses to promote cycling must dedicate considerable time and financial resources to developing suitable infrastructure that guarantees safety while also supporting efficient travel (Aldred and Jungnickel 2014). This protection has been provided by riding lanes and bike-sharing schemes, which constantly prioritize cyclists as the centre of traffic on roadways in these cycling-obsessed cities. Figure 7.9 looks at how the Netherlands, Germany, Kenya, and Uganda have made bicycling among women a feasible, albeit uncomfortable, and practical mode of transportation in their cities.

Local area plans offer enormous potential to focus on smaller areas and conduct in-depth research on these selected areas. At the local level, accessibility is critical in encouraging people to choose sustainable means of transportation. Walking, cycling, and taking public transportation are the most environmentally friendly ways of transportation. Walking and cycling are also good for your health, minimize pollution, and save money on gas and maintenance. The principles of land use transportation assimilation, as well as sustainable transportation in general, have gained traction in the previous 10 to 15 years. With substantial improvements in bus-based transit in all JnNURM cities, the JnNURM offered a much-needed boost to urban public transportation. In India, the authority for city land use planning lies under the purview of the municipal body or development authority of that city. Separate components of transportation planning are overseen by different authorities at the municipal, state, and national levels.

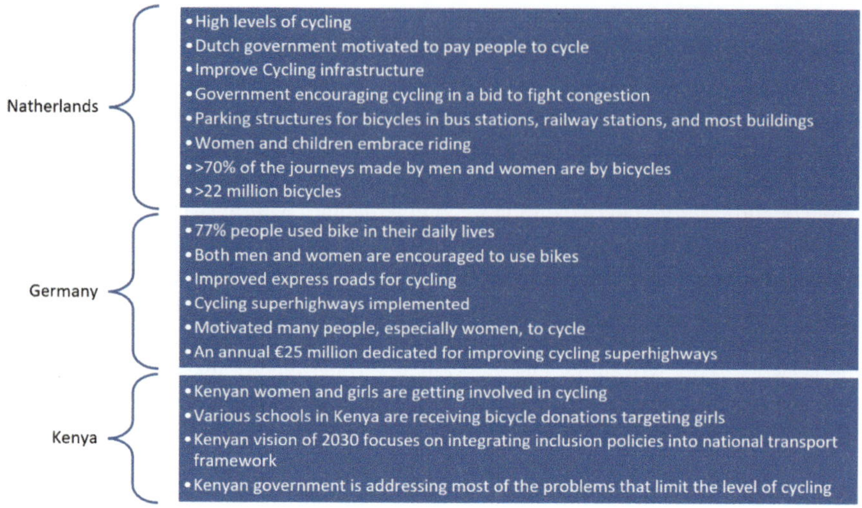

Fig. 7.9 Case studies of women in bicycle transport

7.2.9 Developing Cities Without Slums

Slums are commonly connected with severe environmental issues as well as various forms of social and economic discrimination that prevent people (Mani et al. 2013) and cities from reaching their full potential for human development and economic prosperity (Corburn and Karanja 2014). Thus, exploiting the opportunity for sustainable development generated by urbanization and facilitated by increased local action is critically dependent on the creation of "cities without slums." Nowadays, architecture is attempting to address the concerns of slums and informal settlements. Slums are becoming a common sight in the undeveloped nations, as the rural migration phenomena have delayed immigrant population integration through urban development. According to the UN-Habitat State of the World's Cities (2010) global evaluation of slums, 828 million people, or 33% of the urban population of emerging economies, live in slums. Currently, around 4 billion people reside in metropolitan areas globally, with approximately 1 billion of them living in slums. According to UN-Habitat, by 2050, 6.4 billion people would be living in cities, with 3 billion living in slums if no meaningful framework for action is put in place to tackle the concerns (Yanyan et al. 2020). Furthermore, assessments of slum populations vary greatly among cities due to discrepancies in accounting criteria. For example, it is thought that more than half of the people in cities like Mumbai and Delhi (in India) live in slums, although the 2011 Indian census put the percentages at 41.3% and 14.6%, respectively. Slums have been presented as examples of institutional weaknesses in housing policy, utility services, local administration, and secure tenure (Brelsford et al. 2017). Over the last five decades, African governments have implemented a variety of initiatives to address the issue of informal settlements. Forcible eviction and destruction; resettlement or evacuation; slum upgrading programmes; and, most frequently, the introduction of facilitating methods is among these measures. Target 11 of the Millennium Development Goals now includes the target of "Cities Without Slums." Progress towards this target will be measured using two indicators: (i) the per cent of people who have access to adequate sanitation, and (ii) the percentage of people who have access to secure tenure.

The new Sustainable Development Goals are far more ambitious in terms of ambitions for the next two decades. Furthermore, significant projects for informal settlement renovation and resettlement are being developed at the national level. To satisfy worldwide development targets over the next few decades, it is anticipated that infrastructure investments in developing countries will need to reach $1 trillion per year, with the majority concentrated in disadvantaged regions of developing cities. Slum upgrading is a critical method for attaining these objectives (Mitlin and Satterthwaite 2013), with infrastructure costs contributing for around half of the total. New types of open-source mapping data, such as OpenStreetMap, are becoming more common as data sources for large-scale comparative studies of urban form. Even as parcel-level data becomes more frequent, the OpenStreetMap community has yet to reach an agreement on whether it should be included on the platform.

Case Study: Spatial Approach to Improve Slum Infrastructure in Muslim Para, Kethardanga Basti in Bankura Municipality

The urban community's own living patterns are greatly impacted by the city's dynamic growth. A result of the urban population boom brought on by urbanization is the appearance of labour clumping in Bankura Municipality's major cities. The issue of urban settlements is one of the effects of the rapid urbanization of cities. If the city is not effectively managed, the quality of the current communities will decline as more people move into it. Through the use of Geographic Information Systems (GIS) and remote sensing technologies, this study seeks to pinpoint the services of slum regions in the Bankura Municipality in West Bengal (India). The total estimated slum is calculated as 8.43 sq km.

In order to assure water in the urban slum region, this study looks at a dependable water supply and stormwater services in Bankura City using the capability of GIS. Here, a generic model for the study and definition of GIS needs is offered as the technique. As a result, it entailed a number of steps, beginning with data gathering and analysis and ending with associated maintenance. The GPS is a satellite navigation system that is located in space and allows users to pinpoint the precise location of a planned distribution network's facilities in the actual world. To identify road service facilities and delineate pipeline interventions, stormwater drainage, a house-to-house survey is carried out. The QGIS programme creates a line layer and converts GIS data to the WGS1984 coordinate system.

Using near me analysis, calculates the distance and additional proximity information between road network and the closest building and households. The estimated road accessibility area is calculated as 4.22 km^2, i.e. 50.06% of the entire study area (Fig. 7.10).

Fig. 7.10 Road accessibility within the slum area

In order to see and quantify the accessibility, we evaluated the service areas. Network analysis software can be used to determine if a service area can be reached within a certain radius or to calculate the trip time from one or more facilities. This enables evaluation of service homes inside a network, such as determining where the population of slums can use a facility for stormwater management or water delivery. The water supply services area is covered by 2.68 km^2, i.e. 31.79% of the entire study area. Moreover, the estimated stormwater services network is 1.45 km^2, i.e. 17.20% of the entire study area (Figs. 7.11 and 7.12).

Fig. 7.11 Water supply services area within the slum region

Fig. 7.12 Stormwater services area within the slum region

7.3 Sustainable Urban Innovation and Geospatial Revolution

We are in the midst of a data revolution! Nowadays, more data is being generated than ever before, and it comes from a wide range of sources. Geospatial monitoring, citizen-generated and crowd-sourced data, and big data are becoming more readily available in real time, and they are progressively complementing official statistics. This growing diversity of data is being used by the corporate sector, academia, and civil society to earn money, contributed extensively, drive innovation, and promote advocacy.

GIS technology has had a significant impact on local government over the last four decades, enhancing fundamental record keeping and data administration as well as streamlining a wide range of geospatially related operations, including mapping. The notion of decentralized business systems produced by interconnecting departmental systems started to emerge over the last quarter, largely attributable to the introduction of faster processing and networks. For continuously merging data from numerous distributed databases, this idea relied on services-oriented architecture. Cities are progressively implementing patterns of distributed data mashups (both tabular and map). This aids in the unification and integration of data from a variety of sources within the organization and outside. The web and web services make this new work pattern possible. All local government data will be serverized, transforming GIS and making it easier to deploy in order to make cities smarter. This framework enables GIS professionals to quickly mash together data and construct apps that would otherwise take a lot of time and effort. A services platform will increasingly be used by citizens to interact with government services. Web services that give access to remote data sources will be used to assist the so-called e-government revolution. These services will be the foundation for a new set of capabilities for doing tabular, statistical, geospatial, and visualization analytics.

7.3.1 Integrating Land Use and Transportation

Land use and transportation are inextricably linked. While urban structure impacts transportation supply, transportation supply promotes urban structure. The location and design of spatial development have a significant impact on traffic patterns. At the same time, the location and attributes of major transportation infrastructure have an impact on land use distribution from both a macro and a site plan standpoint. Population, land use and urban structure, and mobility (transportation) behaviour all play a role in the development of urban sustainability. The notion of land use transport integration is based on the relationship between spatial and transportation development. The present situation to spatial and transportation planning is not connected, and transportation planning takes a reactive approach to giving solutions for transportation challenge manifestations such as overcrowding, delays, pollution, and so on. Instead of analysing the potential implications of a specific development pattern, transportation plans take spatial plan recommendations as provided.

Sustainability cannot be achieved solely through urban structure initiatives or sectoral approaches such as transportation infrastructure legislation. Effective land use, transportation, and environmental solutions are sought accordingly. City structure and transportation have impacted each other, with transportation modes/technology determining the design and scale of the city. Historically, cities around the world arose organically and were centred on work centres. The cities were compact, with the central market square and institutional buildings, as well as the dispersal of economic operations along important routes. All attractions have to be within easy reach for cities to be accessible by walk or cycle, resulting in tiny and dense city growth. People's mobility grew with the introduction of automobiles, giving them the ability to live in a better living environment, further away from the city centre (Fig. 7.13).

As the number of private vehicles on the road increased, the city expanded outward. Cities grew in size, and as automobile volumes increased, new broader

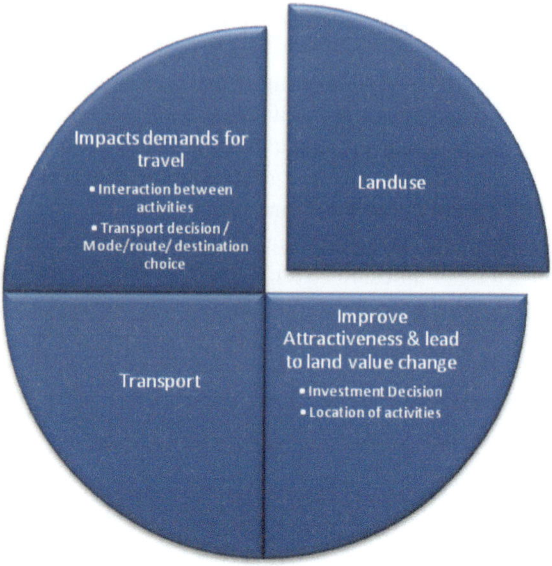

Fig. 7.13 Integration of transport and land use system

roads were built to provide better mobility to diverse destinations to fulfil everyday demands. Thus, the steady increase motorization is resulting in sprawling cities with low density, longer travel distances due to increasing city sizes and dispersed development patterns, and has contributed in traffic congestions and greenhouse gas emissions, worsening of the environment, enhanced accidents, and a decline in urban life standards.

7.3.2 Strategic Spatial Planning and Infrastructure Planning

The term "infrastructure" also implies that surface water, green areas, and stormwater management are equally important as other facilities in the urban context (Lennon 2015). The addition of "blue" to the Green Index (GI)-concept is widespread, and it emphasizes the importance of water in green structure development and administration. They range from an inadequate knowledge about the amount of current Blue-Green Index (BGI), maintenance expenses, and citizens' opinions and desired solutions to technical challenges. Database structure, format of accessible GIS data, patchy data provision due to a lack of data, lack of communication between data gathering bodies at various geographical and temporal dimensions, and insufficient knowledge of how to use the systems are all examples of the latter (Ahern et al. 2014).

To keep up with the support structure of this infrastructure, professionals frequently turn to GIS technologies to address many of the geographical discrepancies that develop within cities. GIS is used in a variety of industries, including utilities, engineering, transportation, logistics, renewables, and maritime. Geographic Information Systems (GIS) aren't just for planning and designing infrastructure. Building and maintenance expenditures may be identified and monitored centrally thanks to asset mapping's integration with project management and budgeting capabilities in GIS. Administrators can use a GIS-based maintenance management system to track and analyse preventive maintenance, which promotes efficient activity budgeting and reporting of work tasks, employees, equipment, and inventory costs. Field staff can collect information, do inspections, and locate assets all at the same time using GIS-enabled mobile devices. The planning, routing, and joining of bridge construction in 3D is made easier by pre-defined structures in 3D modelling tools. 3D innovations aid in the creation of synchronized and constructible designs and construction phases, leading to a better overview of the existing state of the project. These innovations help to improve the quality of construction projects.

The concept "spatial data infrastructure" (SDI) is often used to describe to the basic collection of technologies, rules, and organizational structures that allow geographic data to be available and accessible. The US National Research Council defined spatial data infrastructure in 1993 as a framework of technologies, institutional structures, and policies that can be utilized to make accessing, creating, and transferring geographic data information easier. It includes geographic data and features, as well as supporting documentation, a method for identifying, visualizing, and analysing the data, and a method for making the data accessible. To be functional, SDI must have the organizational arrangements required to govern and administrate it on a local, regional, national, and/or worldwide scale (Fig. 7.14).

SDI is a service that connects different data sources in SDI architecture. This paradigm primarily focuses on SDI elements while also facilitating access to geospatial data, management, and services (Hu and Li 2017). Big Data can be used in a variety of ways in smart cities. Within a few years, massive amounts of data have been collected, and IoT technologies have become a significant role in smart city decision-making. The evolution of the Internet of Things (IOT) offers a variety of services, including sensor integration and the use of Bluetooth in the real world with highly networked services (Al Nuaimi et al. 2015). The suggested definition of big data, which includes parameters such as volume, velocity, and variety, is not only about generating data from different sources, but also about assisting in the realization of sustainability, natural resource intelligence, citizen quality of life, adaptability, and other amenities. It's amazing news that the government has begun to emphasis on smart city ideas in attempt to optimize inhabitants' living standards and quality of life. The Internet of Things has recently been employed in the creation of smart city initiatives in order to build a link and make use of existing resources. This strategy significantly improves the odds of providing clients with a high quality of life (Kumara and Jailia 2018). This is the technology that integrates a significant number of various end systems and is built to assist the goal of a smart city. This technology is required to deliver advanced communication technologies to facilitate added-value services in smart city development initiatives.

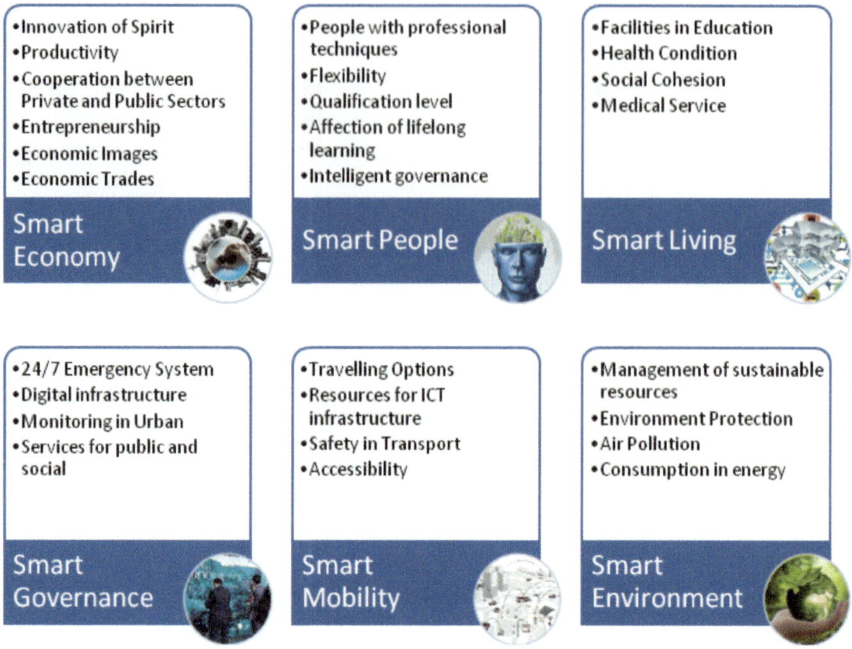

Fig. 7.14 Spatial data infrastructure framework for sustainable city

7.3.3 Integrated Urban Management and Development Plans

Integrated urban governance is a must-have if cities around the world are to meet the various issues they face today. Worldwide, major cities compete predominantly for innovative technology and services, international financial institutions, and places to live for highly trained specialist and managerial professionals. As a result, cities are evolving into a new type of "global player." Simultaneously, they are becoming progressively vulnerable to global forces, which present them with new, difficult challenges and may restrict their capacity to intervene. Not only are favourable circumstances in terms of transportation, land prices, the pool of technical employees, and other "hard" location factors determining how international investors act, but so are "soft" aspects including quality of life, education and culture assistance, as well as crime rates and other aspects. In the majority of areas, migration, rising economic disparity (particularly in the northern hemisphere), a speedily growing middle class in emerging markets and in the southern hemisphere, coupled—in some cases—with tremendous spatial and population growth, are all intensifying social area discrepancies (Fig. 7.15).

For example, the OECD (1996) uses the term "coherent policy making," the WHO and UN ECE (2006) use the term "policy integration," and the UK Cabinet Office (2000) uses the term "crosscutting policy making." Individual authors, on the other hand, have coined their own notions. "Policy coordination" (Challis et al.

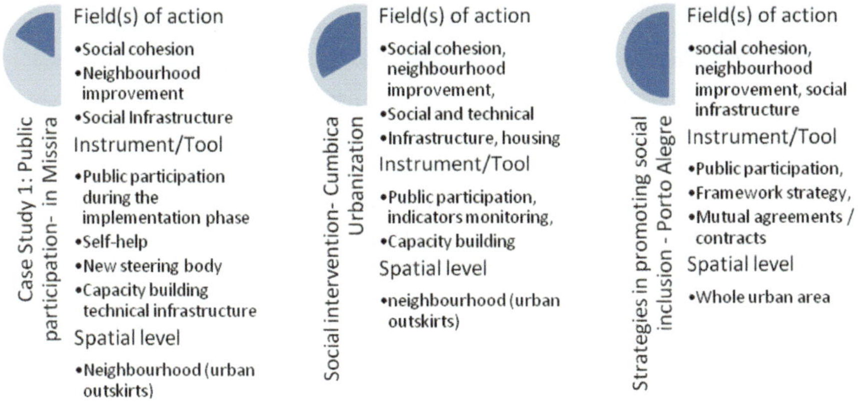

Fig. 7.15 Case studies for integrated urban management and development plan

1988), "concerted decision making" (Warren et al. 1974), "joined up policy" or "joined-up governance" (Wilkinson and Appelbee 1999; Ling 2002), and "policy integration" (Ling 2002) are only a few examples (Stead and de Jong 2006). However, the Integrated Urban Governance refers to:

- Collaboration between municipal authorities' numerous individual departments.
- Government and institutions at different administrative levels and authorities (e.g. district or borough—municipality—region—country) will work together.
- In order to attain (overarching) policy goals, political control is necessary.
- In municipal authority bodies, new decision-making frameworks and/or institutional reforms are being implemented.
- Involving or incorporating civil society and/or business in decision-making and/or execution.
- Comprehensive political techniques that pay more attention to the underlying causes of crises as well as the living conditions of the people.

As a result, Integrated Urban Governance is fundamentally a management strategy. It is concerned with the handling of cross-cutting issues in policymaking that span across recognized policy sectors. Management of policy responsibilities inside a particular organization or sector is also included. Integrated governance encompasses both horizontal and vertical inter-governmental unification (between various levels of government), along with assimilation beyond territorial area (in the double sense of city authorities—regional/national level administration and administration—civil society) (Fig. 7.16).

Fig. 7.16 Case studies of integrated urban development programme

7.4 Digital Sand Model and Sustainable Smart City

The Digital Sand Model Technology combines four major inventions: 3D graphics, high-resolution satellite data, simulation, and Geographic Information Systems (GIS). The system involves the GeorbIS 3D GIS platform to gather data on various terrains, as well as audio and video gear. Armed personnel benefit from enhanced situational awareness because to this innovation. For planning and operations, mission briefings, and training, the Indian army, paramilitary, police organizations, and special forces employ this integrated approach. For operations planning, the digital model provides full 3D visualization of all terrains, encompassing urban, rural, mountainous, and desert environments. It is also feasible to integrate organization-specific workflows for training and input data integration for geo-tagged incident management. It's also feasible to create a 3D scenario for mission briefing that includes a comprehensive fly-through and walkthrough of the operational region, as well as interactive 3D models showing events and unit movement.

Urban planning combines GIS and other infrastructure platforms to combine urban information resources and uses three-dimensional (3D) visualized technology (Liu 2020) and urban digital twin (Dembski et al. 2020) to help design. However, these tools are still displayed on a screen. Users are having a harder time adapting to the new interactive mode, despite the fact that it allows them to control the model more freely. Users still prefer the old sand table interaction method, where they may engage with the model instinctively to better comprehend urban planning and architecture. The emergence of virtual reality (VR) presents an opportunity to re-establish that connectivity. The conventional physical sand table is a model formed of sand and other materials based on a percentage of a topographic map, field landscape, or aerial map. It has the properties of a strong three-dimensional sensation, enabling multiple consumers to feel instantly from any angle while also engaging in certain interactions. Because the size is commonly large, the communication technique is through direct contact with the actual model and moving them around using devices. Despite the short interaction, it is consistent with people's physiological and psychological routines, allowing for effective communication.

We can accomplish the planning in the virtual mapping first, thanks to digital twin technology and the thingjs 3D visualization application framework for the Internet of Things. The digital model of the city is created through the collecting of large data. Create a virtual and real-time possible interaction between the physical and digital cities. IoT energy monitoring approach allows for real-time monitoring of energy consumption by appliances in a facility. The building's administrators can utilize this information to optimize the use of electricity and lower electricity expenses. The digital twin city can exhibit the detailed implementation predicament of the city in all disciplines in a single picture and reveal it by classification according to various topics, so as to realize management fields including implementation monitoring, management, processing, and decision-making of some important scenes, with the aid of smart large screens, urban instrument panels, leadership cockpits, digital sand tables, stereo projection, and other aspects. Detect traffic conditions in real time, analyse traffic congestion, and dredge vehicles in a timely manner. Obtain vehicle road coordination and reduce urban traffic congestion. Track the road, examine vehicle situations at various locations, and track the status of the urban road network's operation and maintenance in real time. The visual blessing allows to intuitively perceive the amount of generation units, the current month's electricity generation capacity, and the comparison between the power generation plan and the fulfilment degree, providing a significant amount of data for the smart electricity approach (Fig. 7.17).

The desire for a better living is represented in many elements of life, including clothing, food, housing, and transportation, as well as education, medical assistance, and occupation. Strengthening the city's "service" quality and increasing people's sense of belonging has become a necessity for urban growth. Despite the fact that there are a thousand different types of smart cities in the viewpoint of a thousand different people, the people-oriented direction is the same. The importance of digital twinning is that it allows things that are highly complex and expensive to do in the real world to be implemented in the virtual world at a low cost and in a short amount of time. "Rome was not constructed in a day," however. With the support of thingjs visualization technology, the smart city requires a step-by-step procedure to better explore the possibilities.

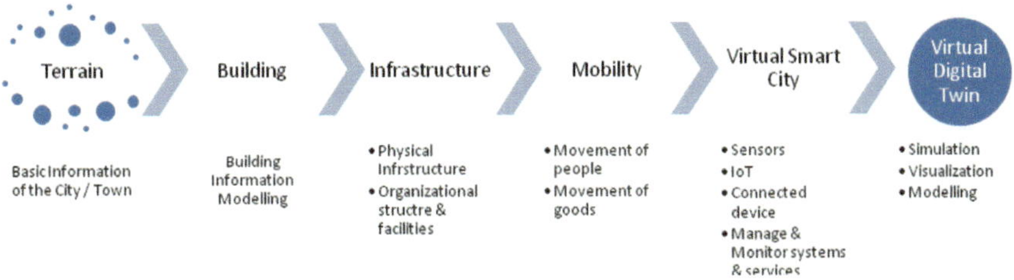

Fig. 7.17 Digital sand model for sustainable smart city

7.5 Way Forward

Millions of people who live in cities today and in the future have a once-in-a-generation opportunity to better their lives because to the digital revolution. But not all populations would profit equally from the rapid adoption of new technology. It is important to connect inclusive growth with smart city development, as this ensures that technological progress benefits all sections of society. National governments play a key role by fostering innovative solutions, investing in human capital development, and supporting policies and programs that strengthen and scale up smart city initiatives. The challenge of scale and information about multiple methods of measurement are two challenges in appropriately evaluating smart city projects and critical characteristics that influence effective measurement (benchmarking, self-assessment, etc.). To accommodate to quickly changing urban surroundings, business and contract models must be more agile from a governance standpoint. Additionally, it is crucial to match the tools to the communication style and include stakeholders at every stage of the policy cycle, considering the range of citizen engagement techniques at one's disposal.

Cities were viewed as the most significant players, but regional and, in particular, national governments were also highly regarded. It was acknowledged that all levels of government lack the competence to solve the problems of establishing a smart city and that they must collaborate to address them. The academic/research community and the financial industry were both evaluated as high priority and low priority, resulting in a divergence of viewpoints. It was recommended that the private sector be divided into categories based on the size and type of businesses. Participants suggested a variety of initiatives, including start-ups and accelerators, the creative/artistic sector (such as start-ups and cultural industries), clusters, supranational business organizations and intercontinental or macro intergovernmental organizations, think tanks and consultancy firms, central governments of developing nations, participants from specific sectors (such as energy, water, and waste water), public financial agencies, players controlling finance, blended finance, and citizens.

Data has been dubbed the "new gold" or "new oil," but it's worth will be limited unless public–private partnerships adapt to a new business environment of smart cities. Long-term contracts known as public–private partnerships (PPPs) include the government and a private partner, wherein the latter shares risk and uses a capital asset to fund and provide public services. PPPs are able to supply public services for social and infrastructure assets (like roads and bridges) as well as facilities like jails, healthcare facilities, and utilities. Using these creative finance models and leveraging synergies is a critical component in assisting urban stakeholders in implementing more efficient governance frameworks through smart city initiatives. Re-regulation, rather than de-regulation, is sometimes required. Technology businesses, in particular, frequently control a considerable portion of their markets, raising the question of whether they are monopolies with the intent to damage consumers. Additionally, in places where digital business models compete with traditional company models, regulation is inconsistent. On the one hand, entrants may claim that laws and regulations created for old market practices are being applied in an unsuitable way to newly formed business models. Traditional market participants benefit from a gap in laws and regulations for innovative business models, providing them an unfair edge. New risk sharing and financing arrangements, as well as more inventive contracts, necessitate regulatory changes.

Another area where urban government will need to change as a result of digital innovation is the support provided to small and medium businesses (SMEs). In the shift to smart, SMEs may encounter both possibilities and obstacles. SME's can use digital technology to increase their market intelligence, achieve scale without mass, and gain access to global markets and knowledge networks at a minimal cost. By means of improved production processes and innovative products or services, the digital era presents SMEs with fresh opportunities to increase their competitiveness in both domestic and global markets. In addition, governments are depending more and more on crowd-sourced data to facilitate prompt responses and acquire accurate real-time information on infrastructure requirements and public service delivery. For example, individuals in many cities can use smartphone applications to notify and alert city officials about potholes, faulty traffic lights, stray rubbish, and other urban issues.

7.6 Conclusion

The efficiency of incorporating big data and managing our possessions for society's use will determine the future expansion of smart cities. Because of the significance of our atmosphere and its reliance on ecosystem facilities, it must be accomplished in a sustainable manner. Effective environmental management will necessitate appropriate assessment, monitoring, response, and strategy invention, all of which can only be accomplished with useful data. Numerous international, national, and municipal authorities or agencies can deliver evidence on the environment of smart cities. Because the amount of data generated will be large, processing it will necessitate technological competence. Prior to constructing smart city infrastructure, data centres must be established, which will aid in the detection of problems as well as the provision of remedies. An active surveillance system of alterations to land use, water quality, green space, and energy consumption should be made available to the public online in order to ensure that society as a whole is eligible for participation.

There are a number of technological and environmental impacts that stand in the way of the smart city mission's implementation, including a lack of policies and guidelines that are required for each smart city. The first problem is to explicitly describe currently devalued environmental mechanisms such as topography, climate, and green space, and to include them in the smart city mission's criteria for identifying smart cities. Second, choosing how best to use resources and make future plans should be a top priority in advancing the creation of smart cities, given the escalating dangers associated with climate change and natural disasters. Because the chunk of people will live in cities in the future, the government must plan for equitable resource allocation in order to encourage long-term growth. This project would need the creation of real-time geographical and time-series information as well as the integration of the requirements outlined in this book as a crucial part of smart city management. The success of the Smart City Mission depends on the integration of the digital ecosystem—which comprises big data, ICT, and IoT—within our natural surroundings.

References

Ahern J, Cilliers S, Niemelä J (2014) The concept of ecosystem services in adaptive urban planning and design: a framework for supporting innovation. Landsc Urban Plann 125:254–259. https://doi.org/10.1016/j.landurbplan.2014.01.020

Albert S (2019) Smart cities: the promises and failures of Utopian technological planning. Available at: phys.org/news

Aldred R, Jungnickel K (2014) Why culture matters for transport policy: the case of cycling in the UK. J Transp Geogr 34:78–87

Al Nuaimi E, Al Neyadi H, Mohamed N, Al-Jaroodi J (2015) Applications of Bigdata to smart cities. J Internet Serv Appl 6(1):252015

Almalki FA, Alsamhi SH, Sahal R et al (2021) Green IoT for eco-friendly and sustainable smart cities: future directions and opportunities. Mobile NetwAppl. https://doi.org/10.1007/s11036-021-01790-w

Alsamhi SH, Ansari MS, Zhao L, Van SN, Gupta SK, Alammari AA, Saber AH, Hebah MYAM, Alasali MAA, Aljabali HM (2019) Tethered balloon technology for green communication in smart cities and healthy environment. In: First international conference of intelligent computing and engineering (ICOICE). IEEE, conference proceedings, pp 1–7

Bamwesigye D, Hlavackova P (2019) Analysis of sustainable transport for smart cities. Sustainability 11(7):2140. https://doi.org/10.3390/su11072140

Benedict MA, McMahon ET (2002) Green infrastructure: smart conservation for the 21st century. Renew Resources J 20(3):12–17

Bibri SE (2020) Advances in the leading paradigms of urbanism and their amalgamation: compact cities, eco–cities, and data–driven smart cities, vol 2020. Springer Nature Switzerland AG, Cham

Brandt N, Nordström L (2011) Stockholm royal seaport: urban smart grid pre-study—final report WP5. Stockholm: Stockholm Royal Seaport / KTH (project report)

Brelsford C, Lobo J, Hand J, Bettencourt LMA (2017) Heterogeneity and scale of sustainable development in cities. Proc Natl Acad Sci U.S.A. 114:8963–8968

Brown G, Raymond CM, Corcoran J (2015) Mapping and measuring place attachment. Appl Geogr 57:42–53

Challis L et al (1988) Joint approaches to social policy: rationality and practice. Cambridge Department for Communites and Local Government (2008): Participation and policy integration in spatial planning. London (http://www.communities.gov.uk/publications/planningandbuilding/Participationintegrationspatial)

Cohen B, Amorós JE (2014) Municipal demand-side policy tools and the strategic management of technology life cycles. Technovation 34(12):797–806

Committee on Climate Change (Adaptation Sub-Committee) (2017) UK Climate Change Risk Assessment 2017: Synthesis report. London: CCC

Copeland E (2014) Small pieces loosely joined, how smarter use of technology and data can deliver real reform of local government. Policy Exchange, January 26, 2015. Retrieved 14 Dec 2016 from http://www.policyexchange.org.uk/publications/category/item/small-pieces-loosely-joined-how-smarter-use-of-technology-and-data-can-deliver-real-reform-of-local-government

Corburn J, Karanja I (2014) Informal settlements and a relational view of health in Nairobi, Kenya: sanitation, gender and dignity. Health Promot Int 31:258–269

Cresswell T (ed) Place a short introduction, vol 58. Blackwell Publishing, Victoria, Australia

Dembski F, Wössner U, Letzgus M, Ruddat M, Yamu C (2020) Urban Digital Twins for smart cities and citizens: the case study of Herrenberg, Germany. Sustainability 12:2307

Dia H (2018) Intelligent mobility for smart cities. Available online: http://www.roads.org.au/Portals/3/CONFERENCE/Hussein_Dia.pdf. Accessed on 22 Oct 2018

Foster L, Heller B, Williams A, Dunn M, Curtis D, Goodwill S (2016) Development of smart inner-city recreational facilities to encourage active living. In: Ubiquitous computing and ambient intelligence. Springer, pp 458–468

Flowers M (2013) Beyond open data: the data-driven city. Data-smart city solutions: an initiative by the Ash Centre at Harvard Kennedy School and powered by Bloomberg Philanthropies. Retrieved 12 Dec 2015, from http://datasmart.ash.harvard.edu/news/article/beyond-open-data-the-data-driven-city-325

Gant J (2015) Smart cities, data warehouses, data lakes and the information management challenge. LG CNS Blog. Retrieved from http://www.lgcnsblog.com/inside-it/smart-cities-data-warehouses-data-lakes-and-the-information-management-challenge

Gray L (2011) World's forests absorb almost 40 percent of man-made CO_2. Available at: telegraph.co.uk/news/earth/earthnews/8708979/

Hitachi Consulting (2015) Hitachi Consulting awarded the opportunity to build big data platform for Copenhagen. Retrieved 12 Dec 2016 from http://www.hitachiconsulting.com/pages/Description.aspx?artcileid=5

Holmstedt L, Brandt N, Robert KH (2017) Can Stockholm royal seaport be part of the puzzle towards global sustainability? From local to global sustainability using the same set of criteria. J Cleaner Prod 140:72–80

Hu Y, Li W (2017) Spatial data infrastructures. arXiv preprint arXiv:1707.03969

Hu MC, Wu CY, Shih T (2015) Creating a new socio-technical regime in China: evidence from the Sino-Singapore Tianjin Eco-City. Futures 70:1–12

Hu MC, Wadin JL, Lo HC, Huang JY (2016) Transformation toward an eco-city: lessons from three Asian cities. J Clean Prod 123:77–87

IBM (2013) IBM's smarter cities challenge. Copenhagen report. Retrieved 9 Jan 2016 from http://smartercitieschallenge.org/scc/executive_reports/SCC-Copenhagen-Report.pdf

Jenkins A, Croitoru A, Crooks AT, Stefanidis A (2016) Crowdsourcing a collective sense of place. PLoS ONE 11:1–20

Jones CB, Purves RS, Clough PD, Joho H (2008) Modelling vague places with knowledge from the web. Int J Geogr Inf Sci 22:1045–1065

Jorgensen BS, Stedman RC (2001) Sense of place as an attitude: lakeshore owners attitudes toward their properties. J Environ Psychol 21:233–248

Jorgensen BS, Stedman RC (2011) Measuring the spatial component of sense of place: a methodology for research on the spatial dynamics of psychological experiences of places. Environ Plan B Plan des 38:795–813

Komninos N (2016) Smart environments and smart growth: connecting innovation strategies and digital growth strategies. Int J Knowl Based Dev 7(3):240–263

Kourtit K (2021) City intelligence for enhancing urban performance value: a conceptual study on data decomposition in smart cities. Asia-Pac J Reg Sci 5:191–222. https://doi.org/10.1007/s41685-021-00193-9

Kramers A, Höjer M, Lövehagen N, Wangel J (2014) Smart sustainable cities–exploring ICT solutions for reduced energy use in cities. Environ Model Softw 56:52–62

Kumara S, Jailia M (2018) Smart cities with spatial data infrastructure and Big Data—a critical review. Int J Adv Stud Scien Res 3(11). Available at SSRN: https://ssrn.com/abstract=3319275

Lazaroiu GC, Roscia M (2012) Definition methodology for the smart cities model. Energy 47(1):326–332
Lennon M (2015) Green infrastructure and planning policy: a critical assessment. Local Environ 20:957–980. https://doi.org/10.1080/13549839.2014.880411
Ling T (2002) Delivering joined-up government in the UK: dimensions, issues and problems. Public Admin 80(4):615ff
Lussault M (2007) L'Homme Spatial. La Construction Sociale de l'espaceHumain: La Construction Sociale de l'espaceHumain. Le Seuil, Paris, France
Lee JH, Hancock MG, Hu MC (2014) Towards an effective framework for building smart cities: lessons from Seoul and San Francisco. Technol Forecast Soc Chang 89:80–99
Liquete C, Kleeschulte S, Dige G, Maes J, Grizzetti B, Olah B, Zulian G (2015) Mapping green infrastructure based on ecosystem services and ecological networks: a Pan-European case study. Environ Sci Policy 54:268–280
Liu C (2018) Oil-rich Abu Dhabi's Masdar city: green oasis or green ghost town? https://www.scmp.com/week-asia/business/article/2133409/oil-rich-abu-dhabis-masdar-city-green-oasis-or-green-ghost-town
Liu X (2020) Three-dimensional visualized urban landscape planning and design based on virtual reality technology. IEEE Access 8:149510–149521
Lowery DR, Morse WC (2013) A qualitative method for collecting spatial data on important places for recreation, livelihoods, and ecological meanings: integrating focus groups with public participation geographic information systems. Soc Nat Resour 26:1422–1437
Marsal-Llacuna ML, Segal ME (2016) The Intelligenter Method (I) for making "smarter" city projects and plans. Cities 55:127–138
Mani A, Mullainathan S, Shafir E, Zhao J (2013) Poverty impedes cognitive function. Science 341:976–980
Massey D (1994) Space, place, and gender. University of Minnesota Press, Minneapolis, MN, USA
Minerva R, Biru A, Rotondi D (2015) Towards a definition of the Internet of Things (IoT). IEEE Internet Initiative 1(1):1–86
Mitlin DC, Satterthwaite D (2013) Urban poverty in the global South: scale and nature. Routledge
Mora L, Bolici R (2016). The development process of smart city strategies: the case of Barcelona. In: Rajaniemi J (ed) Re-city: future city—combining disciplines Juvenes print
NGICP and IGICP (2018). National green infrastructure certification program glossary. Available at: www.ngicp.org/
OECD (1996) Building policy coherence: tools and tensions. OECD Public Management Occasional Paper No 12, Paris
Online Post (2013, May 20) City turns to data to meet climate ambitions. Retrieved 12 Dec 2016 from http://cphpost.dk/news14/national-news14/city-turns-to-data-to-meet-climate-ambitions.html
Olaverri C (2018) Intelligent technologies for mobility in smart cities. Available online: https://www.hte.hu/documents/10180/1727937/HT_2016-1_MJIK2015_4_Olaverri-Monreal.pdf
Perera C, Zaslavsky A, Christen P, Georgakopoulos D (2014) Context aware computing for the internet of things: a survey. IEEE Commun Surv Tutor 16(1):414–454
Powell RA, Mitchell MS (2012) What is a home range? J Mammal 93:948–958
Rapoport E, Vernay AL (2011) Defining the eco–city: a discursive approach. Paper presented at the management and innovation for a sustainable built environment conference, international eco–cities initiative, The Netherlands, Amsterdam, pp 1–15
Roche S (2016) Geographic information science II. Prog Hum Geogr 40:565–573
Relph E (1976) Place and placelessness, vol 67. Pion Ltd., London, UK
SmartCitiesWorld & Philips Lighting (2018) Smart Cities—Understanding the Challenges
Stead D, de Jong M (2006) Supportive institutional conditions for the integration of transport, environment and health issues in policymaking (UN Economic and Social Council) n.p. Stockholm Environment Institute—PINTS—Policy Integration for Sustainability—Background Paper (2004): Environmental Policy Integration: An Introduction. Stockholm(http://www.sei-international.net/mediamanager/documents-/Publications/Policy-institutions/pints_intro.pdf)

Stockholm City (2009) Stockholm royal seaport: vision 2030, https://international.stockholm.se/globalassets/ovriga-bilder-och-filer/visionsrs2030_medium.pdf
Stockholm City (2017) Smart and connected city Stockholm. http://international.stockholm.se/globalassets/ovriga-bilder-och-filer/smart-city/brochure-smart-and-connected.pdf
Stockholm City (2018) https://vaxer.stockholm/globalassets/tema/oversiktplan-ny_light/english_stockholm_city_plan.pdf
Syed F, Gupta SK, Hamood Alsamhi S, Rashid M, Liu X (2020) A survey on recent optimal techniques for securing unmanned aerial vehicles applications. Trans Emerg Telecommun Technol:e4133
Tan S, Yang J, Yan J, Lee C, Hashim H, Chen B (2017) A holistic low carbon city indicator framework for sustainable development. Appl Energy 185:1919–1930
Tsolakis N, Anthopoulos L (2015) Eco-cities: an integrated system dynamics framework and a concise research taxonomy. Sustain Cities Soc 17:1–14
Trindade EP, Hinnig MPF, da Costa EM et al (2017) Sustainable development of smart cities: a systematic review of the literature. J. Open Innov 3:11. https://doi.org/10.1186/s40852-017-0063-2
UK Green Building Council (2015) Green building council demystifying green infrastructure. London
UK Cabinet Office (2000) Wiring it up. Whitehall's management of cross-cutting policies and services. London. http://www.cabinetoffice.gov.uk/media/cabinetoffice/-strategy/assets/coiwire.pdf
UN Habitat (2011) Hot cities: battle-ground for climate change. Global report on human settlement
UNECE, WHO (2006) Practical guidance on institutional arrangements for integrated policy and decision making. Geneva
UN-Habitat State of the World's Cities (2010/2011) Bridging the urban divide. Earthscan London
Vanclay F (2008) Place matters. In: Vanclay F, Matthew H, Blackshaw A (eds) Making sense of place: exploring concepts and expressions of place through different senses and lenses. National Museum of Australia Press, Canberra, Australia, pp 3–11
Wammes H (2015) How spatial big data underpins smart cities. GIS Lounge. Retrieved 2 Jan 2015 from http://www.gislounge.com/how-spatial-big-data-underpins-smart-cities
Warren RL, Rose SM, Bergunder AF (1974) The structure of urban reform. Lexingtons
Wilkinson D, Appelbee E (1999) Implementing Holistic Government: Joined-up Action on the Ground. Bristol: Policy Press. ISBN: 978-1-86134-143-3
Woetzel J, Remes J, Boland B, Lv K, Sinha S, Strube G et al. (2018) Smart cities: digital solutions for a more livable future. McKinsey Global Institute
Wong T-C, Yuen B (2011) Understanding the origins and evolution of eco-city development: an introduction. In: Wong T-C, Yuen B (eds) Eco-city planning: policies, practice and design. Springer, Dordrecht, London, pp 1–16
Yanyan X, Olmos LE, Abbar S, González MC (2020) Deconstructing laws of accessibility and facility distribution in cities. Sci Adv 6:37
Yigitcanlar T, Kamruzzaman M (2015) Planning, development and management of sustainable cities: a commentary from the guest editors. Sustainability 7(11):14677–14688
Yigitcanlar T, Lee SH (2014) Korean ubiquitous-eco-city: a smart-sustainable urban form or a branding hoax? Technol Forecast Soc Chang 89:100–114
Zhou J (2012) Sustainable transportation in the US: a review of proposals, policies, and programs since 2000. Front. Archit. Res. 1:150–165
Zhou N, He G, Williams C, Fridley D (2015) Elite cities: a low-carbon eco-city evaluation tool for China. Ecol Ind 48:448–456

GPSR Compliance

The European Union's (EU) General Product Safety Regulation (GPSR) is a set of rules that requires consumer products to be safe and our obligations to ensure this.

If you have any concerns about our products, you can contact us on ProductSafety@springernature.com

In case Publisher is established outside the EU, the EU authorized representative is:

Springer Nature Customer Service Center GmbH
Europaplatz 3
69115 Heidelberg, Germany

Batch number: 08929720

Printed by Printforce, the Netherlands